Lecture Notes in Computer Science 9479

Commenced Publication in 1973
Founding and Former Series Editors:
Gerhard Goos, Juris Hartmanis, and Jan van Leeuwen

More information about this series at http://www.springer.com/series/7407

David F. Gleich · Júlia Komjáthy
Nelly Litvak (Eds.)

Algorithms and Models for the Web Graph

12th International Workshop, WAW 2015
Eindhoven, The Netherlands, December 10–11, 2015
Proceedings

 Springer

Editors
David F. Gleich
Purdue University
West Lafayette, IN
USA

Nelly Litvak
University of Twente
Enschede
The Netherlands

Júlia Komjáthy
Eindhoven University of Technology
Eindhoven
The Netherlands

ISSN 0302-9743   ISSN 1611-3349   (electronic)
Lecture Notes in Computer Science
ISBN 978-3-319-26783-8   ISBN 978-3-319-26784-5   (eBook)
DOI 10.1007/978-3-319-26784-5

Library of Congress Control Number: 2015954997

LNCS Sublibrary: SL1 – Theoretical Computer Science and General Issues

Springer Cham Heidelberg New York Dordrecht London

Springer International Publishing AG Switzerland is part of Springer Science+Business Media (www.springer.com)

Preface

This volume contains the papers presented at WAW2015, the 12th Workshop on Algorithms and Models for the Web-Graph held during December 10–11, 2015, in Eindhoven.

There were 24 submissions. Each submission was reviewed by at least one, and on average two, Program Committee members. The committee decided to accept 15 papers. The program also included three invited talks, by Mariana Olvera-Cravioto (Columbia University), Remco van der Hofstad (Eindhoven University of Technology), and Paul Van Dooren (Catholic University of Louvain). This year the workshop was accompanied by a school aimed at PhD students, postdocs, and young researchers. The speakers of the school were Dean Eckles (Facebook), David F. Gleich, Kyle Kloster (Purdue University), and Tobias Müller (Utrecht University).

Analyzing data as graphs has transitioned from a minor subfield into a major industrial effort over the past 20 years. The World Wide Web was responsible for much of this growth and the Workshop on Algorithms and Models for the Web-Graph (WAW) originally started by trying to understand the behavior and processes underlying the Web. It has since outgrown these roots and WAW is now one of the premier venues for original research work that blends rigorous theory and experiments in analyzing data as a graph. We believe that the 12th WAW continues the high standards of the earlier workshops and as a result maintains the tradition of a small, high-quality workshop.

The organizers would like to thank EURANDOM, the NETWORKS grant, Microsoft Research, and Google for contributing to the financial aspect of the workshop. We would especially like to thank EURANDOM and the Eindhoven University of Technology for their hospitality and smooth organization of the material aspects of the conference such as drinks/food, accommodation for speakers, etc.

The editorial aspects of the proceedings were supported via the online tool EasyChair. It made our work easier and smoother.

September 2015

David F. Gleich
Júlia Komjáthy
Nelly Litvak
Yana Volkovich

Organization

Program Committee

Konstantin Avrachenkov	Inria Sophia Antipolis, France
Paolo Boldi	Università degli Studi di Milano, Italy
Anthony Bonato	Ryerson University, USA
Colin Cooper	King's College London, UK
Debora Donato	StumbleUpon Inc., USA
Andrzej Dudek	Western Michigan University, USA
Alan Frieze	Carnegie Mellon University, USA
David F. Gleich	Purdue University, USA
Jeannette Janssen	Dalhousie University, Canada
Júlia Komjáthy	Eindhoven University of Technology, The Netherlands
Evangelos Kranakis	Carleton University, Canada
Gábor Kun	Eötvös Loránd University, Hungary
Silvio Lattanzi	Google, New York, USA
Marc Lelarge	Inria-ENS, France
Stefano Leonardi	University of Rome La Sapienza, Italy
Nelly Litvak	University of Twente, The Netherlands
Oliver Mason	National University of Ireland, Maynooth, Ireland
Tobias Mueller	Utrecht University, The Netherlands
Animesh Mukherjee	Indian Institute of Technology, Kharagpur, India
Peter Mörters	University of Bath, UK
Mariana Olvera-Cravioto	Columbia University, NY, USA
Liudmila Ostroumova Prokhorenkova	Yandex, Russia
Pan Peng	TU Dortmund, Germany
Mason Porter	University of Oxford, UK
Paweł Prałat	Ryerson University, USA
Sergei Vassilvitskii	Google, New York, USA
Yana Volkovich	Barcelona Media – Innovation Centre, Spain
Stephen J. Young	University of California, San Diego, USA

Additional Reviewers

Bradonjic, Milan
Kadavankandy, Arun

Contents

Properties of PageRank on Large Graphs

Properties of Large Graph Models

Robustness of Spatial Preferential Attachment Networks

Emmanuel Jacob[1] and Peter Mörters[2(✉)]

[1] École Normale Supérieure de Lyon, Lyon, France
[2] University of Bath, Bath, UK
maspm@bath.ac.uk

Abstract. We study robustness under random attack for a class of networks, in which new nodes are given a spatial position and connect to existing vertices with a probability favouring short spatial distances and high degrees. In this model of a scale-free network with clustering one can independently tune the power law exponent $\tau > 2$ of the degree distribution and a parameter $\delta > 1$ determining the decay rate of the probability of long edges. We argue that the network is robust if $\tau < 2 + \frac{1}{\delta}$, but fails to be robust if $\tau > 2 + \frac{1}{\delta-1}$. Hence robustness depends not only on the power-law exponent but also on the clustering features of the network.

Keywords: Scale-free network · Barabasi-Albert model · Preferential attachment · Geometric random graph · Power law · Clustering · Robustness · Giant component · Resilience

1 Introduction

Scientific, technological or social systems can often be described as complex networks of interacting components. Many of these networks have been empirically found to have strikingly similar topologies, shared features being that they are *scale-free*, i.e. the degree distribution follows a power law, *small worlds*, i.e. the typical distance of nodes is logarithmic or doubly logarithmic in the network size, or *robust*, i.e. the network topology is qualitatively unchanged if an arbitrarily large proportion of nodes chosen at random is removed from the network. Barabási and Albert [2] therefore concluded fifteen years ago 'that the development of large networks is governed by robust self-organizing phenomena that go beyond the particulars of the individual systems.' They suggested a model of a growing family of graphs, in which new vertices are added successively and connected to vertices in the existing graph with a probability proportional to their degree, and a few years later these features were rigorously verified in the work of Bollobás and Riordan, see [5,6,8].

A characteristic feature present in most real networks that is not picked up by preferential attachment is that of *clustering*, the formation of clusters of nodes with an edge density significantly higher than in the overall network. A natural way to integrate this feature in the model is by giving every node an individual

© Springer International Publishing Switzerland 2015
D.F. Gleich et al. (Eds.): WAW 2015, LNCS 9479, pp. 3–14, 2015.
DOI: 10.1007/978-3-319-26784-5_1

feature and implementing a preference for edges connecting vertices with similar features. This is usually done by spatial positioning of nodes and rewarding short edges, see for example [1, 17, 21]. Here we investigate a model, introduced in [19], which is a generalisation of the model of Aiello et al. [1]. It is defined as a growing family of graphs in which a new vertex gets a randomly allocated spatial position on the torus. This vertex then connects to every vertex in the existing graph independently, with a probability which is a decreasing function of the spatial distance of the vertices, the time, and the inverse of the degree of the vertex. The relevance of this *spatial preferential attachment model* lies in the fact that, while it is still a scale-free network governed by a simple rule of self-organisation, it has been shown to exhibit clustering. The present paper investigates the problem of robustness.

In mathematical terms, we call a growing family of graphs *robust* if the critical parameter for vertex percolation is zero, which means that whenever vertices are deleted independently at random from the graph with a positive retention probability, a connected component comprising an asymptotically positive proportion of vertices remains. For several scale-free models, including non-spatial preferential attachment networks, it has been shown that the transition between robust and non-robust behaviour occurs when the power law exponent τ crosses the value three, see for example [5, 14]. Robustness in scale-free networks relies on the presence of a hierarchically organised core of vertices with extremely high degrees, such that every vertex is connected to the next higher layer by a small number of edges, see for example [22]. Our analysis of the spatial model shows that, if $\tau < 3$, whether vertices in the core are sufficiently close in the graph distance to the next higher layer depends critically on the speed at which the connection probability decreases with spatial distance, and hence depending on this speed robustness may hold or fail. The phase transition between robustness and non-robustness therefore occurs at value of τ strictly smaller than three.

The main structural difference between the spatial and classical model of preferential attachment is that the former exhibits *clustering*. Mathematically this is measured in terms of a positive clustering coefficient, meaning that, starting from a randomly chosen vertex, and following two different edges, the probability that the two end vertices of these edges are connected remains positive as the graph size is growing. This implies in particular that local neighbourhoods of typical vertices in the spatial network do not look like trees. However, the main ingredient in almost every mathematical analysis of scale-free networks so far has been the approximation of these neighbourhoods by suitable random trees, see [4, 7, 13, 16]. As a result, the analysis of spatial preferential attachment models requires a range of entirely new methods, which allow to study the robustness of networks without relying on the local tree structure that turned out to be so useful in the past.

2   The Model

While spatial preferential attachment models may be defined in a variety of metric spaces, we focus here on homogeneous space represented by a one-dimensional torus of unit volume, given as $\mathbb{T}_1 = (-1/2, 1/2]$ with the endpoints identified. We use d_1 to denote the torus metric. Let $\mathcal{X}$ denote a homogeneous Poisson point process of finite intensity $\lambda > 0$ on $\mathbb{T}_1 \times (0, \infty)$. A point $\mathbf{x} = (x, s)$ in $\mathcal{X}$ is a vertex $\mathbf{x}$, born at time s and placed at position x. Observe that, almost surely, two points of $\mathcal{X}$ neither have the same birth time nor the same position. We say that (x, s) is *older* than (y, t) if $s < t$. For $t > 0$, write $\mathcal{X}_t$ for $\mathcal{X} \cap (\mathbb{T}_1 \times (0, t])$, the set of vertices already born at time t.

We construct a growing sequence of graphs $(G_t)_{t>0}$, starting from the empty graph, and adding successively the vertices in $\mathcal{X}$ when they are born, so that the vertex set of G_t equals $\mathcal{X}_t$. Given the graph G_{t-} at the time of birth of a vertex $\mathbf{y} = (y, t)$, we connect $\mathbf{y}$, independently of everything else, to each vertex $\mathbf{x} = (x, s) \in G_{t-}$, with probability

$$\varphi\left(\frac{t}{f(Z(\mathbf{x}, t-))} d_1(x, y)\right), \tag{1}$$

where $Z(\mathbf{x}, t-)$ is the *indegree* of vertex $\mathbf{x}$, defined as the total number of edges between $\mathbf{x}$ and younger vertices, at time $t-$. The model parameters in (1) are the *attachment rule* $f: \mathbb{N} \cup \{0\} \rightarrow (0, \infty)$, which is a nondecreasing function regulating the strength of the preferential attachment, and the *profile function* $\varphi: [0, \infty) \rightarrow (0, 1)$, which is an integrable nonincreasing function regulating the decay of the connection probability in terms of the interpoint distance. The connection probabilities in (1) may look arcane at a first glance, but are in fact completely natural. To ensure that the probability of a new vertex connecting to its nearest neighbour does not degenerate, as $t \uparrow \infty$, it is necessary to scale $d_1(x, y)$ by $1/t$, which is the order of the distance of a point to its nearest neighbour at time t. The linear dependence of the argument of φ on time ensures that the expected number of edges connecting a new vertex to vertices of bounded degree remains bounded from zero and infinity, as $t \uparrow \infty$, as long as $x \mapsto \varphi(|x|)$ is integrable.

The model parameters λ, f and φ are not independent. If $\int \varphi(|x|) \, dx = \mu > 0$, we can modify φ to $\varphi \circ (\mu \, \mathrm{Id})$ and f to μf, so that the connection probabilities remain unchanged and

$$\int \varphi(|x|) \, dx = 1. \tag{2}$$

Similarly, if the intensity of the Poisson point process $\mathcal{X}$ is $\lambda > 0$, we can replace $\mathcal{X}$ by $\{(x, \lambda s): (x, s) \in \mathcal{X}\}$ and f by λf, so that again the connection probabilities are unchanged and we get a Poisson point process of unit intensity. From now on we will assume that both of these normalisation conventions are in place. Under these assumptions the regime for the attachment rule f which leads to power law degree distributions is characterised by asymptotic linearity, i.e.

$$\lim_{k \uparrow \infty} \frac{f(k)}{k} = \gamma,$$

for some $\gamma > 0$. We henceforth assume asymptotic linearity with the additional constraint that $\gamma < 1$, which excludes cases with infinite mean degrees.

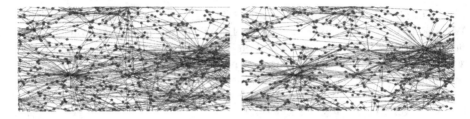

Fig. 1. Simulations of the network for the two-dimensional torus, based on the same realisation of the Poisson process, with parameters $\gamma = 0.75$ and $\delta = 2.5$ (left) and $\delta = 5$ (right). Both networks have the same edge density, but the one with larger δ shows more pronounced clustering. The pictures zoom into a typical part of the torus.

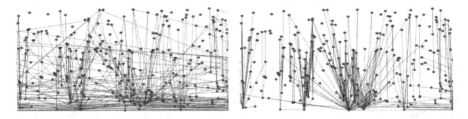

Fig. 2. Simulations of the network for the one-dimensional torus, the vertical axis indicating birth time of the nodes. Parameters are $\gamma = 0.75$ and $\delta = 2$ (left), resp. $\delta = 5$ (right) and both networks have the same edge density and power law exponent. Our results show that the network on the left is robust, the one on the right is not.

We finally assume that the profile function φ is either regularly varying at infinity with index $-\delta$, for some $\delta > 1$, or φ decays quicker than any regularly varying function. In the latter case we set $\delta = \infty$. Intuitively, the bigger δ, the stronger the clustering in the network. See Figs. 1 and 2 for simulations of the spatial preferential attachment network indicative of the parameter dependence.

A similar spatial preferential attachment model was introduced in [1] and studied further in [10,20]. There it is assumed that the profile functions has bounded support, more precisely $\varphi = p\mathbb{1}_{[0,r]}$, for $p \in (0, 1]$ and r satisfying (2). This choice, roughly corresponding to the boundary case $\delta \uparrow \infty$, is too restrictive for the problems we study in this paper, as it turns out that robustness does not hold for any value of τ. Other spatial models with a phase transition between a robust and a non-robust phase are the scale-free percolation model of Deijfen et al. [11], and the Chung-Lu model in hyperbolic space, discussed in Candellero and Fountoulakis [9]. In both cases the transition happens when the power law exponent of the degree distribution crosses the value 3.

Local properties of the spatial preferential attachment model were studied in [19], where this model was first introduced. It is shown there that

- The *empirical degree distribution* of G_t converges in probability to a deterministic limit μ. The probability measure μ on $\{0\} \cup \mathbb{N}$ satisfies

$$\mu(k) = k^{-(1+\frac{1}{\gamma})+o(1)} \qquad \text{as} \quad k \uparrow \infty.$$

The network $(G_t)_{t>0}$ is *scale-free* with power-law exponent $\tau = 1 + \frac{1}{\gamma}$, which can be tuned to take any value $\tau > 2$. See [19, Theorems 1 and 2].

- The average over all vertices $v \in G_t$ of the empirical local clustering coefficient at v, defined as the proportion of pairs of neighbours of v which are themselves connected by an edge in G_t, converges in probability to a positive constant $c_\infty^{\mathrm{av}} > 0$, called the *average clustering coefficient*. In other words the network $(G_t)_{t>0}$ exhibits *clustering*. See [19, Theorem 3].

3 Statement of the Result

Recall that the number of vertices of the graphs G_t, $t > 0$, form a Poisson process of unit intensity, and is therefore almost surely equivalent to t as $t \uparrow \infty$. Let $C_t \subset G_t$ be the largest connected component in G_t and denote by $|C_t|$ its size. We say that the network has a *giant component* if C_t is of linear size or, more precisely, if

$$\lim_{\varepsilon \downarrow 0} \limsup_{t \to \infty} \mathbb{P}\left(\frac{|C_t|}{t} \le \varepsilon \right) = 0;$$

and it has *no giant component* if C_t has sublinear size or, more precisely, if

$$\liminf_{t \to \infty} \mathbb{P}\left(\frac{|C_t|}{t} \le \varepsilon \right) = 1 \text{ for any } \varepsilon > 0.$$

If G is a graph with vertex set $\mathcal{X}$, and $p \in (0,1)$, we write ^pG for the random subgraph of G obtained by Bernoulli percolation with retention parameter p on the vertices of G. We also use $^p\mathcal{X}$ for set of vertices surviving percolation. The network $(G_t)_{t>0}$ is said to be *robust* if, for any fixed $p \in (0,1]$, the network $(^pG_t)_{t>0}$ has a giant component and *non-robust* if there exists $p \in (0,1]$ so that $(^pG_t)_{t>0}$ has no giant component.

Theorem 1. *The spatial preferential attachment network $(G_t)_{t>0}$ is*

(a) robust if $\gamma > \frac{\delta}{1+\delta}$ or, equivalently, if $\tau < 2 + \frac{1}{\delta}$;
(b) non-robust if $\gamma < \frac{\delta-1}{\delta}$ or, equivalently, if $\tau > 2 + \frac{1}{\delta-1}$.

Remark 1. The network is also non-robust if $\gamma < \frac{1}{2}$ or, equivalently, if $\tau > 3$. But the surprising result here is that for $\delta > 2$ the transition between the two phases occurs at a value strictly below 3. This phenomenon is new and due to the clustering structure in the network. It offers a new perspective on the 'classical' results on network models without clustering.

Remark 2

- We conjecture that the result in (a) is sharp, i.e. nonrobustness occurs if $\gamma < \frac{\delta}{1+\delta}$. If this holds, the critical value for τ equals $2 + \frac{1}{\delta}$. Our proof techniques currently do not allow to prove this.

– Our approach also provides heuristics indicating that in the robust phase $\delta(\tau - 2) < 1$ the typical distances in the robust giant component are asymptotically

$$(4 + o(1)) \frac{\log \log t}{-\log(\delta(\tau - 2))},$$

namely doubly logarithmic, just as in some nonspatial preferential attachment models. The constant coincides with that of the nonspatial models in the limiting case $\delta \downarrow 1$, see [12,15], and goes to infinity as $\delta(\tau - 2) \to 1$. It is an interesting open problem to confirm these heuristics rigorously.

4   Proof Ideas and Strategies

Before describing the strategies of our proofs, we briefly summarise the techniques developed in [19] in order to describe the local neighbourhoods of typical vertices by a limit model.

Canonical Representation. We first describe a canonical representation of our network $(G_t)_{t>0}$. To this end, let $\mathcal{X}$ be a Poisson process of unit intensity on $\mathbb{T}_1 \times (0, \infty)$, and endow the point process $\mathcal{X} \times \mathcal{X}$ with independent marks which are uniformly distributed on $[0, 1]$. We denote these marks by $\mathcal{V}_{\mathbf{x},\mathbf{y}}$ or $\mathcal{V}(\mathbf{x}, \mathbf{y})$, for $\mathbf{x}, \mathbf{y} \in \mathcal{X}$. If $\mathcal{Y} \subset \mathbb{T}_1 \times (0, \infty)$ is a finite set and $\mathcal{W} \colon \mathcal{Y} \times \mathcal{Y} \to [0, 1]$ a map, we define a graph $G^1(\mathcal{Y}, \mathcal{W})$ with vertex set $\mathcal{Y}$ by establishing edges in order of age of the younger endvertex. An edge between $\mathbf{x} = (x, t)$ and $\mathbf{y} = (y, s)$, $t < s$, is present if and only if

$$\mathcal{W}(\mathbf{x}, \mathbf{y}) \leq \varphi \left(\frac{s \, d_1(x, y)}{f(Z(\mathbf{x}, s-))} \right), \tag{3}$$

where $Z(\mathbf{x}, s-)$ is the indegree of $\mathbf{x}$ at time $s-$. A realization of $\mathcal{X}$ and $\mathcal{V}$ then gives rise to the family of graphs $(G_t)_{t>0}$ with vertex sets $\mathcal{X}_t = \mathcal{X} \cap (\mathbb{T}_1 \times (0, t])$, given by $G_t = G^1(\mathcal{X}_t, \mathcal{V})$, which has the distribution of the spatial preferential attachment network.

Space-Time Rescaling. The construction above can be generalised in a straightforward manner from $\mathbb{T}_1$ to the torus of volume t, namely $\mathbb{T}_t = (-\frac{1}{2} t, \frac{1}{2} t]$, equipped with its canonical torus metric d_t. The resulting functional, mapping a finite subset $\mathcal{Y} \subset \mathbb{T}_t \times (0, \infty)$ and a map from $\mathcal{Y} \times \mathcal{Y} \to [0, 1]$ onto a graph, is now denoted by G^t. We introduce the *rescaling mapping*

$$h_t : \mathbb{T}_1 \times (0, t] \to \mathbb{T}_t \times (0, 1],$$
$$(x, s) \quad \mapsto (tx, s/t)$$

which expands the space by a factor t, the time by a factor $1/t$. The mapping h_t operates on the set $\mathcal{X}$, but also on $\mathcal{V}$, by $h_t(\mathcal{V})_{h_t(\mathbf{x}), h_t(\mathbf{y})} := \mathcal{V}_{\mathbf{x},\mathbf{y}}$. The operation of h_t preserves the rule (3), and it is therefore simple to verify that we have

$$G^t(h_t(\mathcal{X}_t), h_t(\mathcal{V})) = h_t(G^1(\mathcal{X}_t, \mathcal{V})) = h_t(G_t),$$

that is, it is the same to construct the graph and then rescale the picture, or to first rescale the picture, then construct the graph on this rescaled picture. Observe also that $h_t(\mathcal{X}_t)$ is a Poisson point process of intensity 1 on $\mathbb{T}_t \times (0,1]$, while $h_t(\mathcal{V})$ are independent marks attached to the points of $h_t(\mathcal{X}_t) \times h_t(\mathcal{X}_t)$ which are uniformly distributed on $[0,1]$.

Convergence to the Limit Model. We now denote by $\mathcal{X}$ a Poisson point process with unit intensity on $\mathbb{R} \times (0,1]$, and endow the points of $\mathcal{X} \times \mathcal{X}$ with independent marks $\mathcal{V}$, which are uniformly distributed on $[0,1]$. For each $t > 0$, identify $(-\frac{1}{2}t, \frac{1}{2}t]$ and $\mathbb{T}_t$, and write $\mathcal{X}^t$ for the restriction of $\mathcal{X}$ to $\mathbb{T}_t \times (0,1]$, and $\mathcal{V}^t$ for the restriction of $\mathcal{V}$ to $\mathcal{X}^t \times \mathcal{X}^t$. In the following, we write G^t or $G^t(\mathcal{X}, \mathcal{V})$ for $G^t(\mathcal{X}^t, \mathcal{V}^t)$. We have seen that for fixed $t \in (0, \infty)$, the graphs G^t and $h_t(G_t)$ have the same law. Thus any results of robustness we prove for the network $(G^t)_{t>0}$ also hold for the network $(G_t)_{t>0}$. It was shown in [19, Proposition 5] that, almost surely, the graphs G^t converge to a locally finite graph $G^\infty = G^\infty(\mathcal{X}, \mathcal{V})$, in the sense that the neighbours of any given vertex $\mathbf{x} \in \mathcal{X}$ coincide in G^t and in G^∞, if t is large enough. It is important to note the fundamentally different behaviour of the processes $(G^t)_{t>0}$ and $(G_t)_{t>0}$. While in the former the degree of any fixed vertex stabilizes, in the latter the degree of any fixed vertex goes to ∞, as $t \uparrow \infty$. We will exploit the convergence of G^t to G^∞ in order to decide the robustness of the finite graphs G^t, and ultimately G_t, from properties of the limit model G^∞.

Law of Large Numbers. We now state a limit theorem for the graphs $^p G^t$ centred in a randomly chosen point. To this end we denote by $^p\mathbb{P}$ the law of $\mathcal{X}, \mathcal{V}$ together with independent Bernoulli percolation with retention parameter p on the points of $\mathcal{X}$. For any $\mathbf{x} \in \mathbb{R} \times (0,1]$ we denote by $^p\mathbb{P}_\mathbf{x}$ the *Palm measure*, i.e. the law $^p\mathbb{P}$ conditioned on the event $\{\mathbf{x} \in {}^p\mathcal{X}\}$. Note that by elementary properties of the Poisson process this conditioning simply adds the point $\mathbf{x}$ to $^p\mathcal{X}$ and independent marks $\mathcal{V}_{\mathbf{x},\mathbf{y}}$ and $\mathcal{V}_{\mathbf{y},\mathbf{x}}$, for all $\mathbf{y} \in \mathcal{X}$, to $\mathcal{V}$. We also write $^p\mathbb{E}_\mathbf{x}$ for the expectation under $^p\mathbb{P}_\mathbf{x}$. Let $\xi = \xi(\mathbf{x}, G)$ be a bounded functional of a locally-finite graph G with vertices in $\mathbb{R} \times (0,1]$ and a vertex $\mathbf{x} \in G$, which is invariant under translations of $\mathbb{R}$. Also, let $\xi_t = \xi_t(\mathbf{x}, G)$ be a bounded family of functionals of a graph G with vertices in $\mathbb{T}_t \times (0,1]$ and a vertex $\mathbf{x} \in G$, invariant under translations of the torus. We assume that, for U an independent uniform random variable on $(0,1]$, we have that $\xi_t((0,U), {}^p G^t)$ converges to $\xi((0,U), {}^p G^\infty)$ in $^p\mathbb{P}_{(0,U)}$-probability. By [19, Theorem 7], in $^p\mathbb{P}$-probability,

$$\frac{1}{t} \sum_{\mathbf{x} \in {}^p\mathcal{X}^t} \xi_t\left(\mathbf{x}, {}^p G^t\right) \xrightarrow[t \to \infty]{} p \int_0^p \mathbb{E}_{(0,u)}[\xi((0,u), {}^p G^\infty)]\, du. \tag{4}$$

4.1 Robustness: Strategy of Proof

Existence of an Infinite Component in the Limit Model. We first show that, under the assumptions that $\gamma > \frac{\delta}{1+\delta}$, or equivalently $\frac{\gamma}{\delta(1-\gamma)} > 1$, the percolated limit model $^P G^\infty$ has an infinite connected component. This uses the established strategy of the hierarchical core. Young vertices, born after time $\frac{1}{2}$, are called *connectors*. We find $\alpha > 1$ such that, starting from a sufficiently old vertex $\mathbf{x}_0 \in {}^P G^\infty$, we establish an infinite chain $(\mathbf{x}_k)_{k \geq 1}$ of vertices $\mathbf{x}_k = (x_k, s_k)$ such that $s_k < s_{k-1}^\alpha$, i.e. we move to increasingly older vertices, and $\mathbf{x}_{k-1}$ and $\mathbf{x}_k$ are connected by a path of length two, using a connector as a stepping stone. The following lemma is the key. Roughly speaking, we call a vertex born at time s *good* if its indegree at time $\frac{1}{2}$ is close to its expectation, i.e. of order $s^{-\gamma}$.

Lemma 1. *Choose first $\alpha \in (1, \frac{\gamma}{\delta(1-\gamma)})$ then $\beta \in (\alpha, \frac{\gamma}{\delta}(1 + \alpha\delta))$. If x is a good vertex born at time s, then with very high probability there exists a good vertex y born before time s^α with $|x - y| < s^{-\beta}$ such that x and y are connected through a connector.*

Proof (Sketch).

- The existence of a good vertex y is easy because it just needs to be located in a box of sidelengths s^α and $2s^{-\beta}$, and $s^\alpha s^{-\beta} \to \infty$.
- At time $\frac{1}{2}$ the good vertex x has indegree of order $s^{-\gamma}$. The number of connectors at distance $\leq s^{-\gamma}$, which are connected to x is therefore stochastically bounded from below by a Poisson variable with intensity $s^{-\gamma}$.
- For each of these connectors the probability that they connect to a good y is at least

$$\varphi\left(\frac{\frac{1}{2}d(x,y)}{s^{-\alpha\gamma}}\right) \leq \mathrm{cst}.s^{-\delta(\alpha\gamma-\beta)}.$$

We succeed because $-\gamma - \delta(\alpha\gamma - \beta) < 0$.

Transfer to Finite Graphs Using the Law of Large Numbers. To infer robustness of the network $(G^t)_{t>0}$ from the behaviour of the limit model we use (4) on the functional $\xi_t(\mathbf{x}, G)$ defined as the indicator of the event that there is a path in G connecting $\mathbf{x}$ to the oldest vertex of G. We denote by $\xi(\mathbf{x}, G)$ the indicator of the event that the connected component of $\mathbf{x}$ is infinite and let

$$^P\theta := \int_0^1 {}^P\mathbb{P}_{(0,u)}\{\text{the component of } (0,u) \text{ in } {}^P G^\infty \text{ is infinite}\}\, du. \qquad (5)$$

If $\lim \xi_t((0,U), {}^P G^t) = \xi((0,U), {}^P G^\infty)$ in probability, then the law of large numbers (4) implies that $\lim(1/t) \sum_{\mathbf{x} \in {}^P \chi^t} \xi_t(\mathbf{x}, {}^P G^t) = p\, {}^P\theta$. The sum is the number of vertices in $^P G^t$ connected to the oldest vertex, and we infer that this number grows linearly in t so that a giant component exists in $({}^P G^t)_{t>0}$. This implies that $(G^t)_{t>0}$ and hence $(G_t)_{t>0}$ is a robust network. However, while it is easy to see that $\limsup_{t \uparrow \infty} \xi_t((0,U), {}^P G^t) \leq \xi((0,U), {}^P G^\infty)$, checking that

$$\liminf_{t \uparrow \infty} \xi_t((0,U), {}^P G^t) \geq \xi((0,U), {}^P G^\infty), \qquad (6)$$

is the difficult part of the argument.

The Geometric Argument. The proof of (6) is the most technical part of the proof. We first look at the finite graph $^pG^t$ and establish the existence of a core of old and well-connected vertices, which includes the oldest vertex. Any pair of vertices in the core are connected by a path with a bounded number of edges, in particular all vertices of the core are in the same connected component. This part of the argument is similar to the construction in the limit model. We then use a simple continuity argument to establish that if the vertex $(0, U)$ is in an infinite component in the limit model, then it is also in an infinite component for the limit model based on a Poisson process $\mathcal{X}$ with a slightly reduced intensity. In the main step we show that under this assumption the vertex $(0, U)$ is connected in $^pG^t$ with reduced intensity to a moderately old vertex. In this step we have to rule out explicitly the possibilities that the infinite component of $^pG^\infty$ either avoids the set of eligible moderately old vertices, or connects to them only by a path which moves very far away from the origin. The latter argument requires good control over the length of edges in the component of $(0, U)$ in $^pG^\infty$. Once the main step is established, we can finally use the still unused vertices, which form a Poisson process with small but positive intensity, to connect the moderately old vertex we have found to the core by means of a classical sprinkling argument.

4.2 Non-robustness: Strategy of Proof

Using the Limit Model. If $\gamma < \frac{1}{2}$ it is very plausible that the spatial preferential attachment network is non-robust, as the classical models with the same power-law exponents are non-robust [5,14] and it is difficult to see how the spatial structure could help robustness. We have not been able to use this argument for a proof, though, as our model cannot be easily dominated by a non-spatial model with the same power-law exponent. Instead we use a direct approach, which turns out to yield non-robustness also in some cases where $\gamma > \frac{1}{2}$. The key is again the use of the limit model, and in particular the law of large numbers. We apply this now to the functionals $\xi^{(k)}(\mathbf{x}, G)$ defined as the indicator of the event that the connected component of $\mathbf{x}$ has no more than k vertices. By the law of large numbers (4) the proportion of vertices in $^pG^t$ which are in components no bigger than k converge, as first $t \uparrow \infty$ and then $k \uparrow \infty$ to $1 - {}^p\theta$. Hence if $^p\theta = 0$ for some $p > 0$, then $(G^t)_{t>0}$ and hence $(G_t)_{t>0}$ is non-robust. It is therefore sufficient to show that, for some sufficiently small $p > 0$, there is no infinite component in the percolated limit model $^pG^\infty$.

Positive Correlation Between Edges. We first explain why a naïve first moment calculation fails. If $(0, U)$ has positive probability of belonging to an infinite component of $^pG^\infty$ then, with positive probability, we could find an infinite self-avoiding path in $^pG^\infty$ starting from $\mathbf{x}_0 = (0, U)$. A direct first moment calculation would require to give a bound on the probability of the event $\{\mathbf{x}_0 \leftrightarrow \mathbf{x}_1 \leftrightarrow \cdots \leftrightarrow \mathbf{x}_n\}$ that a sequence $(\mathbf{x}_0, \ldots, \mathbf{x}_n)$ of distinct points $\mathbf{x}_i = (x_i, s_i)$ conditioned to be in $\mathcal{X}$ forms a path in G^∞. If this estimate allows us to bound the expected number of paths of length n in G^∞ starting in $\mathbf{x}_0 = (0, U)$ by C^n, for some constant C, we can infer with Borel-Cantelli that, if $p < 1/C$, almost surely there is no arbitrarily long self-avoiding paths in $^pG^\infty$. The problem here is that the events $\{x_j \leftrightarrow x_{j+1}\}$ and $\{x_k \leftrightarrow x_{k+1}\}$ are positively correlated if the interval $I = (s_j, s_{j+1}) \cap (s_k, s_{k+1})$ is nonempty, because the existence of a vertex in $\mathcal{X} \cap (\mathbb{R} \times I)$ may make their indegrees grow simultaneously. Because the positive correlations play against us, it seems not possible to give an effective upper bound on the probability of a long sequence to be a path, therefore making this first moment calculation impossible.

Quick Paths, Disjoint Occurrence, and the BK Inequality. As a solution to this problem we develop the concept of *quick paths*. If $^pG^\infty$ contains an infinite path, then there is an infinite quick path in G^∞ with at least half of its points lying in $^pG^\infty$. The expected number of quick paths of length n can be bounded by C^n, for some $C > 0$, and the naïve argument above can be carried through.

Starting with a geodesic path $x_0 \leftrightarrow \cdots \leftrightarrow x_\ell$ in $^pG_0{}^\infty$ we first construct a subsequence $y_n = x_{\varphi(n)}$ by letting $\varphi(0) = 0$ and $\varphi(n+1)$ be the maximal $k > \varphi(n)$ such that there is $y \in G^\infty$ younger than $x_{\varphi(n)}$ and x_k with $x_{\varphi(n)} \leftrightarrow y \leftrightarrow x_k$. We emphasise that y need not be in $^pG^\infty$ but only in G^∞. The vertex y is called a *common child* of the vertices $x_{\varphi(n)}$ and $x_{\varphi(n+1)}$, and if there is no common child we let $\varphi(n+1) = \varphi(n) + 1$. The *quick path* $z_0 \leftrightarrow \cdots \leftrightarrow z_m$ associated with the geodesic path $x_0 \leftrightarrow \cdots \leftrightarrow x_\ell$ is obtained by inserting between y_n and y_{n+1}, if they are not connected by an edge, their oldest common child $y \in G^\infty$. Quick paths are characterised by the properties;

(i) A vertex which is not a local maximum (i.e. younger than its two neighbours in the chain) cannot be connected by an edge to a younger vertex of the path, except possibly its neighbours.

(ii) Two vertices z_n and z_{n+j}, with $j \geq 2$, which are not local maxima, can have common children only if $j = 2$ and z_{n+1} is a local maximum. In that case, z_{n+1} is their oldest common child.

Introduce a *splitting at index i* if either z_i is younger than both z_{i-1} and z_{i-2}, or younger than both z_{i+1} and z_{i+2}. We write $n_0 = 0 < n_1 < \cdots < n_k = m$ for the splitting indices in increasing order. Let

$$A_j = \{z_{n_{j-1}} \leftrightarrow \cdots \leftrightarrow z_{n_j}\}.$$

Then if $z_0 \leftrightarrow \cdots \leftrightarrow z_m$ is a path in G^∞ that satisfies (i) and (ii), then $A_1, \ldots, A_k$ occur disjointly. The concept of disjoint occurrence is due to van den Berg and Kesten. Two increasing events A and B *occur disjointly* if there exists disjoint subsets of the domain of the Poisson process such that A occurs if the points falling in the first subset are present, and B occurs if the points falling in the second subset are present. The famous *BK-inequality*, see [3] for the variant most useful in our context, states that the probability of events occurring disjointly is bounded by the product of their probabilities. The events A_j involve five or fewer consecutive vertices and Fig. 3 shows the six possible types, up to symmetry. The probability of these types can be estimated by a direct calculation.

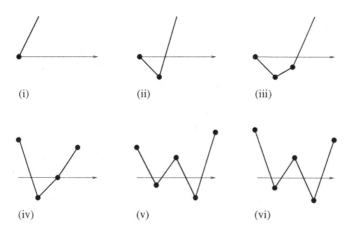

Fig. 3. Up to symmetry there are six types of small parts after the splitting. Illustrated, with the index of a point on the abscissa and time on the ordinate, these are (i) one single edge, (ii) a V shape with two edges, (iii) a V shape with three edges and the end vertex of the short leg between the two vertices of the long leg, (iv) a V shape with three edges and both vertices of the long leg below the end vertex of the short leg, (v) a W shape with the higher end vertex on the side of the deeper valley, (vi) a W shape with the lower end vertex on the side of the deeper valley.

An Refinement of the Method. The method described so far, allows to show non-robustness only in the case $\tau > 3$. To show non-robustness in the case $\tau > 2 + \frac{1}{\delta-1}$ a refinement is needed, which we now briefly describe.

A vertex z born at time u has typically of order $u^{-\gamma}$ younger neighbours, which may be a lot. As most of these neighbours are close to z, namely within distance u^{-1}, and their local neighbourhoods are therefore strongly correlated, our bounds are far from sharp. No matter how many vertices within distance u^{-1} of z belong to the component of z, it will not help much to connect z to vertices far away. Indeed, defining the *region around* z as

$$C_z = \{z' \text{ born at } u' \geq u, |z' - z| \leq 2u^{-1} - u'^{-1}\},$$

we show that the typical number of vertices outside C_z that are connected to z, or any other vertex in C_z, is only of order $\log(u^{-1})$. To estimate the probability of a path it therefore makes sense to take all the points within C_z for granted and consider only those edges of a quick path straddling a suitably defined boundary of C_z. This improves our bounds because few edges straddle the boundary, and the boundary remains small as u becomes small.

Acknowledgements. We gratefully acknowledge support of this project by the *European Science Foundation* through the research network *Random Geometry of Large Interacting Systems and Statistical Physics (RGLIS)*, and by *CNRS*. A full version of this paper has been submitted for publication elsewhere [18].

References

1. Aiello, W., Bonato, A., Cooper, C., Janssen, J., Prałat, P.: A spatial web graph model with local influence regions. Internet Math. **5**, 175–196 (2009)
2. Barabási, A.-L., Albert, R.: Emergence of scaling in random networks. Science **286**, 509–512 (1999)
3. van den Berg, J.: A note on disjoint-occurrence inequalities for marked poisson point processes. J. Appl. Probab. **33**(2), 420–426 (1996)
4. Berger, N., Borgs, C., Chayes, J.T., Saberi, A.: Asymptotic behavior and distributional limits of preferential attachment graphs. Ann. Probab. **42**(1), 1–40 (2014)
5. Bollobás, B., Riordan, O.: Robustness and vulnerability of scale-free random graphs. Internet Math. **1**(1), 1–35 (2003)
6. Bollobás, B., Riordan, O.: The diameter of a scale-free random graph. Combinatorica **24**(1), 5–34 (2004)
7. Bollobás, B., Riordan, O.: Random graphs and branching processes. In: Bollobás, B., Kozma, R., Miklós, D. (eds.) Handbook of Large-Scale Random Networks. Bolyai Society Mathematical Studies, vol. 18, pp. 15–115. Springer, Berlin (2009)
8. Bollobás, B., Riordan, O., Spencer, J., Tusnády, G.: The degree sequence of a scale-free random graph process. Random Struct. Algorithms **18**(3), 279–290 (2001)
9. Candellero, E., Fountoulakis, N.: Bootstrap percolation and the geometry of complex networks, 1–33 (2014). Preprint arXiv:1412.1301
10. Cooper, C., Frieze, A., Prałat, P.: Some typical properties of the spatial preferred attachment model. In: Bonato, A., Janssen, J. (eds.) WAW 2012. LNCS, vol. 7323, pp. 29–40. Springer, Heidelberg (2012)
11. Deijfen, M., van der Hofstad, R., Hooghiemstra, G.: Scale-free percolation. Ann. Inst. Henri Poincaré Probab. Statist. **49**(3), 817–838 (2013)
12. Dereich, S., Mönch, C., Mörters, P.: Typical distances in ultrasmall random networks. Adv. Appl. Probab. **44**(2), 583–601 (2012)
13. Dereich, S., Mörters, P.: Random networks with concave preferential attachment rule. Jahresber. Dtsch. Math.-Ver. **113**(1), 21–40 (2011)
14. Dereich, S., Mörters, P.: Random networks with sublinear preferential attachment: the giant component. Ann. Probab. **41**(1), 329–384 (2013)
15. Dommers, S., van der Hofstad, R., Hooghiemstra, G.: Diameters in preferential attachment models. J. Stat. Phys. **139**(1), 72–107 (2010)
16. Eckhoff, M., Mörters, P.: Vulnerability of robust preferential attachment networks. Electron. J. Probab. **19**(57), 47 (2014)
17. Flaxman, A.D., Frieze, A.M., Vera, J.: A geometric preferential attachment model of networks. Internet Math. **3**(2), 187–205 (2006)
18. Jacob, E., Mörters, P.: Robustness of scale-free spatial networks, 1–34 (2015). Preprint arXiv:1504.00618
19. Jacob, E., Mörters, P.: Spatial preferential attachment: power laws and clustering coefficients. Ann. Appl. Prob. **25**, 632–662 (2015)
20. Janssen, J., Pralat, P., Wilson, R.: Geometric graph properties of the spatial preferred attachment model. Adv. Appl. Math. **50**, 243–267 (2013)
21. Jordan, J.: Geometric preferential attachment in non-uniform metric spaces. Electron. J. Probab. **18**(8), 15 (2013)
22. Norros, I., Reittu, H.: Network models with a 'soft hierarchy': a random graph construction with loglog scalability. IEEE Netw. **22**(2), 40–47 (2008)

Local Clustering Coefficient in Generalized Preferential Attachment Models

Alexander Krot[1]([✉]) and Liudmila Ostroumova Prokhorenkova[2,3]

[1] Moscow Institute of Physics and Technology, Moscow, Russia
al.krot.kav@gmail.com
[2] Yandex, Moscow, Russia
[3] Moscow State University, Moscow, Russia

Abstract. In this paper, we analyze the local clustering coefficient of preferential attachment models. A general approach to preferential attachment was introduced in [19], where a wide class of models (PA-class) was defined in terms of constraints that are sufficient for the study of the degree distribution and the clustering coefficient. It was previously shown that the degree distribution in all models of the PA-class follows a power law. Also, the global clustering coefficient was analyzed and a lower bound for the average local clustering coefficient was obtained. We expand the results of [19] by analyzing the local clustering coefficient for the PA-class of models. Namely, we analyze the behavior of $C(d)$ which is the average local clustering for the vertices of degree d.

Keywords: Networks · Random graph models · Preferential attachment · Clustering coefficient

1 Introduction

Nowadays there are a lot of practical problems connected with the analysis of growing real-world networks, from Internet and society networks [1,6,9] to biological networks [2]. Models of real-world networks are used in physics, information retrieval, data mining, bioinformatics, etc. An extensive review of real-world networks and their applications can be found elsewhere (e.g., see [1,6,7,13]).

It turns out that many real-world networks of diverse nature have some typical properties: small diameter, power-law degree distribution, high clustering, and others [15,17,18,24]. Probably the most extensively studied property of networks is their vertex degree distribution. For the majority of studied real-world networks, the portion of vertices with degree d was observed to decrease as $d^{-\gamma}$, usually with $2 < \gamma < 3$ [3–6,10,14].

Another important characteristic of a network is its clustering coefficient, which has the following two most used versions: the global clustering coefficient and the average local clustering coefficient (see Sect. 2.3 for the definitions). It is believed that for many real-world networks both the average local and the global clustering coefficients tend to non-zero limit as the network becomes large.

© Springer International Publishing Switzerland 2015
D.F. Gleich et al. (Eds.): WAW 2015, LNCS 9479, pp. 15–28, 2015.
DOI: 10.1007/978-3-319-26784-5_2

Indeed, in many observed networks the values of both clustering coefficients are considerably high [18].

The most well-known approach to modeling complex networks is the preferential-attachment idea. Many different models are based on this idea: LCD [8], Buckley-Osthus [11], Holme-Kim [16], RAN [25], and many others. A general approach to preferential attachment was introduced in [19], where a wide class of models was defined in terms of constraints that are sufficient for the study of the degree distribution (PA-class) and the clustering coefficient (T-subclass of PA-class).

In this paper, we analyze the behavior of $C(d)$ — the average local clustering coefficient for the vertices of degree d — in the T-subclass. It was previously shown that in real-world networks $C(d)$ usually decreases as $d^{-\psi}$ with some parameter $\psi > 0$ [12,21,23]. For some networks, $C(d)$ scales as a power law $C(d) \sim d^{-1}$ [13,20]. In the current paper, we prove that in *all* models of the T-subclass the local clustering coefficient $C(d)$ asymptotically behaves as $C \cdot d^{-1}$, where C is some constant.

The remainder of the paper is organized as follows. In Sect. 2, we give a formal definition of the PA-class and present some known results. Then, in Sect. 3, we state new results on the behavior of local clustering $C(d)$. We prove the theorems in Sect. 4. Section 5 concludes the paper.

2 Generalized Preferential Attachment

2.1 Definition of the PA-class

In this section, we define the PA-class of models which was first suggested in [19]. Let G_m^n ($n \geq n_0$) be a graph with n vertices $\{1, \dots, n\}$ and mn edges obtained as a result of the following process. We start at the time n_0 from an arbitrary graph $G_m^{n_0}$ with n_0 vertices and mn_0 edges. On the $(n+1)$-th step ($n \geq n_0$), we make the graph G_m^{n+1} from G_m^n by adding a new vertex $n+1$ and m edges connecting this vertex to some m vertices from the set $\{1, \dots, n, n+1\}$. Denote by d_v^n the degree of a vertex v in G_m^n. If for some constants A and B the following conditions are satisfied

$$\mathsf{P}\left(d_v^{n+1} = d_v^n \mid G_m^n\right) = 1 - A\frac{d_v^n}{n} - B\frac{1}{n} + O\left(\frac{(d_v^n)^2}{n^2}\right), \quad 1 \leq v \leq n, \qquad (1)$$

$$\mathsf{P}\left(d_v^{n+1} = d_v^n + 1 \mid G_m^n\right) = A\frac{d_v^n}{n} + B\frac{1}{n} + O\left(\frac{(d_v^n)^2}{n^2}\right), \quad 1 \leq v \leq n, \qquad (2)$$

$$\mathsf{P}\left(d_v^{n+1} = d_v^n + j \mid G_m^n\right) = O\left(\frac{(d_v^n)^2}{n^2}\right), \quad 2 \leq j \leq m, \ 1 \leq v \leq n, \qquad (3)$$

$$\mathsf{P}(d_{n+1}^{n+1} = m + j) = O\left(\frac{1}{n}\right), \quad 1 \leq j \leq m, \qquad (4)$$

then the random graph process G_m^n is a model from the PA-class. Here, as in [19], we require $2mA + B = m$ and $0 \leq A \leq 1$.

As it is explained in [19], even fixing values of parameters A and m does not specify a concrete procedure for constructing a network. There are a lot of models possessing very different properties and satisfying the conditions (1–4), e.g., the LCD, the Buckley–Osthus, the Holme–Kim, and the RAN models.

2.2   Power Law Degree Distribution

Let $N_n(d)$ be the number of vertices of degree d in G_m^n. The following theorems on the expectation of $N_n(d)$ and its concentration were proved in [19].

Theorem 1. *For every model in PA-class and for every* $d \geq m$

$$\mathsf{E}N_n(d) = c(m,d)\left(n + O\left(d^{2+\frac{1}{A}}\right)\right),$$

where

$$c(m,d) = \frac{\Gamma\left(d + \frac{B}{A}\right)\Gamma\left(m + \frac{B+1}{A}\right)}{A\Gamma\left(d + \frac{B+A+1}{A}\right)\Gamma\left(m + \frac{B}{A}\right)} \overset{d\to\infty}{\sim} \frac{\Gamma\left(m + \frac{B+1}{A}\right)d^{-1-\frac{1}{A}}}{A\Gamma\left(m + \frac{B}{A}\right)}$$

and $\Gamma(x)$ *is the gamma function.*

Theorem 2. *For every model from the PA-class and for every* $d = d(n)$ *we have*

$$\mathsf{P}\left(|N_n(d) - \mathsf{E}N_n(d)| \geq d\sqrt{n}\log n\right) = O\left(n^{-\log n}\right).$$

Therefore, for any $\delta > 0$ *there exists a function* $\varphi(n) \in o(1)$ *such that*

$$\lim_{n\to\infty} \mathsf{P}\left(\exists d \leq n^{\frac{A-\delta}{4A+2}} : |N_n(d) - \mathsf{E}N_n(d)| \geq \varphi(n)\,\mathsf{E}N_n(d)\right) = 0.$$

These two theorems mean that the degree distribution follows (asymptotically) the power law with the parameter $1 + \frac{1}{A}$.

2.3   Clustering Coefficient

A T-subclass of the PA-class was introduced in [19]. In this case, the following additional condition is required:

$$\mathsf{P}\left(d_i^{n+1} = d_i^n + 1, d_j^{n+1} = d_j^n + 1 \mid G_m^n\right) = e_{ij}\frac{D}{mn} + O\left(\frac{d_i^n d_j^n}{n^2}\right). \qquad (5)$$

Here e_{ij} is the number of edges between vertices i and j in G_m^n and D is a positive constant. Note that this property still does not define the correlation between edges completely, but it is sufficient for studying both global and average local clustering coefficients.

Let us now define the clustering coefficients. The *global clustering coefficient* $C_1(G)$ is the ratio of three times the number of triangles to the number of pairs of adjacent edges in G. The *average local clustering coefficient* is defined as follows: $C_2(G) = \frac{1}{n}\sum_{i=1}^{n} C(i)$, where $C(i)$ is the local clustering coefficient for a vertex i: $C(i) = \frac{T^i}{P_2^i}$, where T^i is the number of edges between neighbors of the vertex i and P_2^i is the number of pairs of neighbors. Note that both clustering coefficients are defined for graphs without multiple edges.

The following theorem on the global clustering coefficient in the T-subclass was proven in [19].

Theorem 3. *Let G_m^n belong to the T-subclass with $D > 0$. Then, for any $\varepsilon > 0$*

*(1) If $2A < 1$, then **whp** $\frac{6(1-2A)D-\varepsilon}{m(4(A+B)+m-1)} \le C_1(G_m^n) \le \frac{6(1-2A)D+\varepsilon}{m(4(A+B)+m-1)}$;*

*(2) If $2A = 1$, then **whp** $\frac{6D-\varepsilon}{m(4(A+B)+m-1)\log n} \le C_1(G_m^n) \le \frac{6D+\varepsilon}{m(4(A+B)+m-1)\log n}$;*

*(3) If $2A > 1$, then **whp** $n^{1-2A-\varepsilon} \le C_1(G_m^n) \le n^{1-2A+\varepsilon}$.*

Theorem 3 shows that in some cases ($2A \ge 1$) the global clustering coefficient $C_1(G_m^n)$ tends to zero as the number of vertices grows.

The average local clustering coefficient $C_2(G_m^n)$ was not fully analyzed previously, but it was shown in [19] that $C_2(G_m^n)$ does not tend to zero for the T-subclass with $D > 0$. In the next section, we fully analyze the behavior of the average local clustering coefficient for the vertices of degree d.

3   The Average Local Clustering for the Vertices of Degree d

In this section, we analyze the asymptotic behavior of $C(d)$ — the average local clustering for the vertices of degree d. Let $T_n(d)$ be the number of triangles on the vertices of degree d in G_m^n (i.e., the number of edges between the neighbors of the vertices of degree d). Then, $C(d)$ is defined in the following way:

$$C(d) = \frac{T_n(d)}{N_n(d)\binom{d}{2}}. \tag{6}$$

In other words, $C(d)$ is the local clustering coefficient averaged over all vertices of degree d. In order to estimate $C(d)$ we should first estimate $T_n(d)$. After that, we can use Theorems 1 and 2 on the behavior of $N_n(d)$.

We prove the following result on the expectation of $T_n(d)$.

Theorem 4. *Let G_m^n belong to the T-subclass of the PA-class with $D > 0$. Then*

(1) if $2A < 1$, then $\mathsf{E}T_n(d) = K(d)\left(n + O\left(d^{2+\frac{1}{A}}\right)\right)$;

(2) if $2A = 1$, then $\mathsf{E}T_n(d) = K(d)\left(n + O\left(d^{2+\frac{1}{A}} \cdot \log(n)\right)\right)$;

(3) if $2A > 1$, then $\mathsf{E}T_n(d) = K(d)\left(n + O\left(d^{2+\frac{1}{A}} \cdot n^{2A-1}\right)\right)$;

where $K(d) = c(m,d)\left(D + \frac{D}{m} \cdot \sum_{i=m}^{d-1}\frac{i}{Ai+B}\right) \overset{d\to\infty}{\sim} \frac{D}{Am} \cdot \frac{\Gamma\left(m+\frac{B+1}{A}\right)}{A\Gamma\left(m+\frac{B}{A}\right)} \cdot d^{-\frac{1}{A}}$.

Second, we show that the number of triangles on the vertices of degree d is highly concentrated around its expectation.

Theorem 5. *Let G_m^n belong to the T-subclass of the PA-class with $D > 0$. Then for every $d = d(n)$*

(1) if $2A < 1$: $P\left(|T_n(d) - ET_n(d)| \geq d^2\sqrt{n}\log n\right) = O\left(n^{-\log n}\right)$;
(2) if $2A = 1$: $P\left(|T_n(d) - ET_n(d)| \geq d^2\sqrt{n}\log^2 n\right) = O\left(n^{-\log n}\right)$;
(3) if $2A > 1$: $P\left(|T_n(d) - ET_n(d)| \geq d^2 n^{2A-\frac{1}{2}}\log n\right) = O\left(n^{-\log n}\right)$.

Consequently, for any $\delta > 0$ there exists a function $\varphi(n) = o(1)$ such that

(1) if $2A \leq 1$: $\lim_{n\to\infty} P\left(\exists d \leq n^{\frac{A-\delta}{4A+2}} : |T_n(d) - ET_n(d)| \geq \varphi(n)\,ET_n(d)\right) = 0$;
(2) if $2A > 1$:
$$\lim_{n\to\infty} P\left(\exists d \leq n^{\frac{A(3-4A)-\delta}{4A+2}} : |T_n(d) - ET_n(d)| \geq \varphi(n)\,ET_n(d)\right) = 0.$$

As a consequence of Theorems 1, 2, 4, and 5, we get the following result on the average local clustering coefficient $C(d)$ for the vertices of degree d in G_m^n.

Theorem 6. *Let G_m^n belong to the T-subclass of the PA-class. Then for any $\delta > 0$ there exists a function $\varphi(n) = o(1)$ such that*

(1) if $2A \leq 1$: $\lim_{n\to\infty} P\left(\exists d \leq n^{\frac{A-\delta}{4A+2}} : \left|C(d) - \frac{K(d)}{\binom{d}{2}c(m,d)}\right| \geq \frac{\varphi(n)}{d}\right) = 0$;

(2) if $2A > 1$: $\lim_{n\to\infty} P\left(\exists d \leq n^{\frac{A(3-4A)-\delta}{4A+2}} : \left|C(d) - \frac{K(d)}{\binom{d}{2}c(m,d)}\right| \geq \frac{\varphi(n)}{d}\right) = 0$.

Note that $\frac{K(d)}{\binom{d}{2}c(m,d)} = \frac{2D}{d(d-1)m}\left(m + \sum_{i=m}^{d-1}\frac{i}{Ai+B}\right) \overset{d\to\infty}{\sim} \frac{2D}{mA}\cdot d^{-1}$.

It is important to note that Theorems 5 and 6 are informative only for $A < \frac{3}{4}$, since only in this case the value $n^{\frac{A(3-4A)-\delta}{4A+2}}$ grows.

In the next section, we first prove Theorem 4. Then, using the Azuma–Hoeffding inequality, we prove Theorem 5. Theorem 6 is a corollary of Theorems 1, 2, 4, and 5.

4   Proofs

In all the proofs we use the notation $\theta(\cdot)$ for error terms. By $\theta(X)$ we denote an arbitrary function such that $|\theta(X)| < X$.

4.1   Proof of Theorem 4

We need the following auxiliary theorem.

Theorem 7. *Let W_n be the sum of the squares of the degrees of all vertices in a model from the PA-class. Then*

(1) if $2A < 1$, then $\mathsf{E}W_n = O(n)$,
(2) if $2A = 1$, then $\mathsf{E}W_n = O(n \cdot \log(n))$,
(3) if $2A > 1$, then $\mathsf{E}W_n = O(n^{2A})$.

This statement is mentioned in [19] and it can be proved by induction. Also, let $S(n,d)$ be the sum of the degrees of all the neighbors of all vertices of degree d. Note that $S(n,d)$ is not greater than the sum of the degrees of the neighbors of all vertices. The last is equal to W_n, because each vertex of degree d adds d^2 to the sum of the degrees of the neighbors of all vertices. So, for any d we have

$$\mathsf{E}S(n,d) \le \mathsf{E}W_n. \tag{7}$$

Now we can prove Theorem 4. Note that we do not take into account the multiplicities of edges when we calculate the number of triangles, since the clustering coefficient is defined for graphs without multiple edges. This does not affect the final result since the number of multiple edges is small for graphs constructed according to the model [7].

We prove the statement of Theorem 4 by induction on d. Also, for each d we use induction on n. First, consider the case $d = m$. The expected number of triangles on any vertex t of degree m is equal to $\mathsf{E}\sum_{(i,j)\in E(G_m^t)}\left(e_{ij}\frac{D}{mt} + O\left(\frac{d_i^t d_j^t}{t^2}\right)\right)$ (see (5)). As G_m^t has exactly mt edges, we get $\mathsf{E}\sum_{(i,j)\in E(G_m^t)}\left(e_{ij}\frac{D}{mt} + O\left(\frac{d_i^t d_j^t}{t^2}\right)\right) = D + o(1)$. The fact that $\mathsf{E}\sum_{(i,j)\in E(G_m^t)}O\left(\frac{d_i d_j}{t^2}\right) = O\left(\frac{\mathsf{E}W_t}{t^2}\right) = o(1)$ can be shown by induction using the conditions (1–4). We also know (see Theorem 1) that $\mathsf{E}N_n(m) = c(m,m)\,n + O(1)$. So, $\mathsf{E}T_n(m) = (D + o(1))\,(c(m,m)\,n + O(1)) = K(m)\,(n + O(1))$. This concludes the proof for the case $d = m$ for all values of A ($2A < 1$, $2A = 1$ and $2A > 1$).

Consider the case $d > m$. Note that the number of triangles on a vertex of degree d is $O(d)$, since this number is $O(1)$ when this vertex appears plus at each step we get a triangle only if we hit both the vertex under consideration and a neighbor of this vertex, and our vertex degree equals d, therefore we get at most dm triangles. Also, $\mathsf{E}N_n(d) = c(m,d)\left(n + O\left(d^{2+\frac{1}{A}}\right)\right)$. So we have $\mathsf{E}T_n(d) = O(d)\,c(m,d)\left(n + O\left(d^{2+\frac{1}{A}}\right)\right)$. In particular, for $n \le Q \cdot d^2$ (where the constant Q depends only on A and m and will be defined later) we have $\mathsf{E}T_n(d) = O\left(c(m,d)\,d^{3+\frac{1}{A}}\right) = O\left(d^2\right) = K(d) \cdot O\left(d^{2+\frac{1}{A}}\right)$. This concludes the proof for the case $d > m$, $n \le Qd^2$ for all values of A.

Now, consider the case $d > m$, $n > Q\,d^2$. Once we add a vertex $n+1$ and m edges, we have the following possibilities.

1. At least one edge hits a vertex of degree d. Then $T_n(d)$ is decreased by the number of triangles on this vertex (because this vertex is a vertex of degree $d + 1$ now). The probability to hit a vertex of degree d is $\frac{Ad+B}{n} + O\left(\frac{d^2}{n^2}\right)$.

Summing over all vertices of degree d we obtain that $ET_n(d)$ is decreased by:

$$\left(\frac{Ad + B}{n} + O\left(\frac{d^2}{n^2}\right)\right) \cdot ET_n(d).\tag{8}$$

2. Exactly one edge hits a vertex of degree $d-1$. Then $T_n(d)$ is increased by the number of triangles on this vertex. The probability to hit a vertex of degree $d-1$ once is equal to $\frac{A(d-1)+B}{n} + O\left(\frac{d^2}{n^2}\right)$. Summing over all vertices of degree $d-1$ we obtain that the value $ET_n(d)$ is increased by:

$$\left(\frac{A(d-1) + B}{n} + O\left(\frac{d^2}{n^2}\right)\right) \cdot ET_n(d-1).\tag{9}$$

3. Exactly one edge hits a vertex of degree $d-1$ and another edge hits its neighbor. Then, in addition to (9), $T_n(d)$ is increased by 1. The probability to hit a vertex of degree $d-1$ and its neighbor is equal to $\frac{D}{mn} + O\left(\frac{(d-1)\,d_i}{n^2}\right)$, where d_i is the degree of this neighbor. Summing over the neighbors of a given vertex of degree $d-1$ and summing then over all vertices of degree $d-1$ we obtain that $ET_n(d)$ is increased by:

$$(d-1)\,EN_n(d-1)\,\frac{D}{mn} + O\left(\frac{d \cdot E\sum_{\substack{i:i \text{ is a neighbor} \\ \text{of a vertex of degree } d-1}} d_i}{n^2}\right)$$

$$= (d-1)\,EN_n(d-1)\,\frac{D}{mn} + O\left(\frac{d\,ES(n,d)}{n^2}\right).\tag{10}$$

4. Exactly i edges hit a vertex of degree $d-i$, where i is between 2 and m. If no edges hit the neighbors of this vertex, then $T_n(d)$ is increased only by the number of triangles on this vertex. The probability to hit a vertex of degree $d-i$ exactly i times is equal to $O\left(\frac{d^2}{n^2}\right)$. If we also hit its neighbors, then $T_n(d)$ is additionally increased by 1 for each neighbor. The probability to hit a vertex of degree $d-i$ exactly i times and hit some its neighbor is, obviously, $O\left(\frac{d^2}{n^2}\right)$. Summing over all vertices of degree $d-i$ and then summing over all i from 2 to m, we obtain that $ET_n(d)$ is increased by:

$$\sum_{i=2}^{m}\left(ET_n(d-i) \cdot O\left(\frac{d^2}{n^2}\right) + O\left(\frac{d^2}{n^2}\right) \cdot (d-i) \cdot EN_n(d-i)\right)$$

$$= O\left(\frac{d^2}{n^2}\right) ET_n(d) + O\left(\frac{d^3}{n^2}\right) EN_n(d).\tag{11}$$

Finally, using (8)–(11) and the linearity of the expectation, we get

$$\mathsf{E}T_{n+1}(d) = \mathsf{E}T_n(d) - \left(\frac{Ad + B}{n} + O\left(\frac{d^2}{n^2} \right) \right) \mathsf{E}T_n(d)$$

$$+ \left(\frac{A(d-1) + B}{n} + O\left(\frac{d^2}{n^2} \right) \right) \mathsf{E}T_n(d-1) + (d-1)\,\mathsf{E}N_n(d-1)\,\frac{D}{mn}$$

$$+ O\left(\frac{d\,\mathsf{E}S(n,d)}{n^2} \right) + O\left(\frac{d^2}{n^2} \right) \mathsf{E}T_n(d) + O\left(\frac{d^3}{n^2} \right) \mathsf{E}N_n(d)$$

$$= \left(1 - \frac{Ad + B}{n} \right) \mathsf{E}T_n(d) + \frac{A(d-1) + B}{n}\,\mathsf{E}T_n(d-1)$$

$$+ O\left(\frac{d^2}{n^2} \right) (\mathsf{E}T_n(d) + \mathsf{E}T_n(d-1)) + O\left(\frac{d^3}{n^2} \right) \mathsf{E}N_n(d)$$

$$+ \frac{D}{mn}(d-1)\,\mathsf{E}N_n(d-1) + O\left(\frac{d \cdot \mathsf{E}S(n,d)}{n^2} \right). \tag{12}$$

Consider the case $2A < 1$ (the cases $2A = 1$ and $2A > 1$ will be analyzed similarly). We prove by induction on d and n that

$$\mathsf{E}T_n(d) = K(d)\left(n + \theta\left(C \cdot d^{2+\frac{1}{A}} \right) \right) \tag{13}$$

for some constant $C > 0$. Let us assume that $\mathsf{E}T_i(\tilde{d}) = K(\tilde{d})\left(i + \theta\left(C \cdot \tilde{d}^{2+\frac{1}{A}} \right) \right)$ for $\tilde{d} < d$ and all i and for $\tilde{d} = d$ and $i < n + 1$.

Recall that $K(d) = c(m,d)\left(D + \frac{D}{m} \cdot \sum_{i=m}^{d-1} \frac{i}{Ai+B} \right)$ and $\mathsf{E}N_n(d) = c(m,d) \cdot \left(n + O\left(d^{2+\frac{1}{A}} \right) \right)$. If $2A < 1$, then from (7) and Theorem 7 we get $\mathsf{E}S(n,d) = O(n)$ and we obtain:

$$\mathsf{E}T_{n+1}(d) = \left(1 - \frac{Ad + B}{n} \right) K(d)\left(n + \theta\left(Cd^{2+\frac{1}{A}} \right) \right)$$

$$+ \frac{A(d-1) + B}{n} K(d-1)\left(n + \theta\left(C(d-1)^{2+\frac{1}{A}} \right) \right)$$

$$+ O\left(\frac{d^2}{n^2} \right) \left(K(d)\left(n + \theta\left(Cd^{2+\frac{1}{A}} \right) \right) + K(d-1)\left(n + \theta\left(C(d-1)^{2+\frac{1}{A}} \right) \right) \right)$$

$$+ O\left(\frac{d^3}{n^2} \right) c(m,d)\left(n + O\left(d^{2+\frac{1}{A}} \right) \right)$$

$$+ \frac{D}{mn}(d-1)\,c(m,d-1)\left(n + O\left(d^{2+\frac{1}{A}} \right) \right) + O\left(\frac{d}{n} \right).$$

Note that $K(d) = \frac{A(d-1)+B}{Ad+B+1} K(d-1) + \frac{D(d-1)}{m(Ad+B+1)} c(m,d-1)$. Therefore, we obtain:

$$ET_{n+1}(d) = K(d)(n+1) + K(d)\left(1 - \frac{Ad+B}{n}\right)\theta\left(C\,d^{2+\frac{1}{A}}\right)$$

$$+ K(d-1)\frac{A(d-1)+B}{n}\theta\left(C(d-1)^{2+\frac{1}{A}}\right)$$

$$+ \frac{D(d-1)}{mn}c(m,d)\,O\left(d^{2+\frac{1}{A}}\right) + O\left(\frac{d}{n}\right) + O\left(\frac{d^2}{n^2}\right)(K(d)\,n$$

$$+ K(d)\,\theta\left(C\,d^{2+\frac{1}{A}}\right) + K(d-1)\,n + K(d-1)\,\theta\left(C(d-1)^{2+\frac{1}{A}}\right))$$

$$+ O\left(\frac{d^3}{n^2}\right)\left(c(m,d)\,n + c(m,d)\,O\left(d^{2+\frac{1}{A}}\right)\right).$$

In order to show (13), it remains to prove that for some large enough C:

$$K(d)\left(\frac{Ad+B}{n}\right)C\,d^{2+\frac{1}{A}} \geq K(d-1)\frac{A(d-1)+B}{n}C\,(d-1)^{2+\frac{1}{A}}$$

$$+ O\left(\frac{d^2}{n}\right) + O\left(C\frac{d^4}{n^2}\right) + O\left(\frac{d^4}{n^2}\right). \tag{14}$$

First, we analyze the following difference:

$$K(d)\left(\frac{Ad+B}{n}\right)d^{2+\frac{1}{A}} - K(d-1)\frac{A(d-1)+B}{n}(d-1)^{2+\frac{1}{A}}$$

$$= \frac{Ad+B}{n}d^{2+\frac{1}{A}}\left(\frac{A(d-1)+B}{Ad+B+1}K(d-1) + \frac{D(d-1)}{m(Ad+B+1)}c(m,d-1)\right)$$

$$- \frac{A(d-1)+B}{n}K(d-1)(d-1)^{2+\frac{1}{A}} = \frac{(Ad+B)D(d-1)}{mn(Ad+B+1)}c(m,d-1)\,d^{2+\frac{1}{A}}$$

$$+ K(d-1)\frac{A(d-1)+B}{n}\left(\frac{Ad+B}{Ad+B+1}d^{2+\frac{1}{A}} - (d-1)^{2+\frac{1}{A}}\right)$$

$$\geq \frac{(Ad+B)D(d-1)}{mn(Ad+B+1)}c(m,d-1)\,d^{2+\frac{1}{A}}$$

$$+ (d-1)^{2+\frac{1}{A}}K(d-1)\frac{A(d-1)+B}{n}\cdot\frac{2A^2d+2AB+B}{Ad(Ad+B+1)}$$

$$\geq \frac{(Ad+B)D(d-1)}{mn(Ad+B+1)}c(m,d-1)\,d^{2+\frac{1}{A}}.$$

Therefore, Eq. (14) becomes:

$$C\frac{(Ad+B)D(d-1)}{mn(Ad+B+1)}c(m,d-1)\,d^{2+\frac{1}{A}} \geq O\left(\frac{d^2}{n}\right) + O\left(C\frac{d^4}{n^2}\right) + O\left(\frac{d^4}{n^2}\right).$$

In the case $2A = 1$ this inequality will be:

$$C\frac{(Ad+B)D(d-1)}{mn(Ad+B+1)}c(m,d-1)\,d^{2+\frac{1}{A}}\log(n)$$

$$\geq O\left(\frac{d^2}{n}\right) + O\left(C\frac{d^4\cdot\log(n)}{n^2}\right) + O\left(\frac{d^4}{n^2}\right) + O\left(\frac{d\log(n)}{n}\right).$$

In the case $2A > 1$ this inequality will be:

$$C \frac{(Ad + B)D(d-1)}{mn(Ad + B + 1)} c(m, d-1) d^{2+\frac{1}{A}} n^{2A-1}$$

$$\geq O\left(\frac{d^2}{n}\right) + O\left(C \frac{d^4 n^{2A-1}}{n^2}\right) + O\left(\frac{d^4}{n^2}\right) + O\left(\frac{d\, n^{2A}}{n^2}\right).$$

It is easy to see that for $n \geq Q \cdot d^2$ (for some large Q which depends only on the parameters of the model) these three inequalities are satisfied. This concludes the proof of the theorem.

4.2 Proof of Theorem 5

This theorem is proved similarly to the concentration theorem from [19]. We also need the following notation (introduced in [19]):

$$p_n(d) = \mathsf{P}\left(d_v^{n+1} = d \mid d_v^n = d\right) = 1 - A\frac{d}{n} - B\frac{1}{n} + O\left(\frac{d^2}{n^2}\right),$$

$$p_n^1(d) := \mathsf{P}\left(d_v^{n+1} = d + 1 \mid d_v^n = d\right) = A\frac{d}{n} + B\frac{1}{n} + O\left(\frac{d^2}{n^2}\right),$$

$$p_n^j(d) := \mathsf{P}\left(d_v^{n+1} = d + j \mid d_v^n = d\right) = O\left(\frac{d^2}{n^2}\right), \quad 2 \leq j \leq m,$$

$$p_n := \sum_{k=1}^{m} \mathsf{P}(d_{n+1}^{n+1} = m + k) = O\left(\frac{1}{n}\right).$$

To prove Theorem 5 we also need the Azuma–Hoeffding inequality:

Theorem 8 (Azuma, Hoeffding). *Let* $(X_i)_{i=0}^n$ *be a martingale such that* $|X_i - X_{i-1}| \leq c_i$ *for any* $1 \leq i \leq n$. *Then* $\mathsf{P}\left(|X_n - X_0| \geq x\right) \leq 2e^{-\frac{x^2}{2\sum_{i=1}^n c_i^2}}$ *for any* $x > 0$.

Consider the random variables $X_i(d) = \mathsf{E}(T_n(d) \mid G_m^i)$, $i = 0, \ldots, n$. Note that $X_0(d) = \mathsf{E}T_n(d)$ and $X_n(d) = T_n(d)$. It is easy to see that $X_n(d)$ is a martingale.

We will prove below that for any $i = 0, \ldots, n-1$

(1) if $2A < 1$, then $|X_{i+1}(d) - X_i(d)| \leq Md^2$,
(2) if $2A = 1$, then $|X_{i+1}(d) - X_i(d)| \leq Md^2 \log(n)$,
(3) if $1 < 2A < \frac{3}{2}$, then $|X_{i+1}(d) - X_i(d)| \leq Md^2 n^{2A-1}$,

where $M > 0$ is some constant. The theorem follows from this statement immediately. Indeed, consider the case $2A < 1$. Put $c_i = Md^2$ for all i. Then from Azuma–Hoeffding inequality it follows that

$$\mathsf{P}\left(|T_n(d) - \mathsf{E}T_n(d)| \geq d^2 \sqrt{n} \log n\right) \leq 2\exp\left\{-\frac{n\, d^4 \log^2 n}{2\, n\, M^2 d^4}\right\} = O\left(n^{-\log n}\right).$$

Therefore, for the case $2A < 1$ the first statement of the theorem is satisfied. If $d \leq n^{\frac{A-\delta}{4A+2}}$, then the value $n\,d^{-1/A}$ is considerably greater than $d^2 \log n \sqrt{n}$. From this the second statement of the theorem follows. The cases $2A = 1$ and $2A > 1$ can be considered similarly. It remains to estimate $|X_{i+1}(d) - X_i(d)|$.

Fix $0 \leq i \leq n-1$ and some graph G_m^i. Note that

$$\left| \mathsf{E}\left(T_n(d) \mid G_m^{i+1}\right) - \mathsf{E}\left(T_n(d) \mid G_m^i\right) \right| \leq \max_{\tilde{G}_m^{i+1} \supset G_m^i} \left\{ \mathsf{E}\left(T_n(d) \mid \tilde{G}_m^{i+1}\right) \right\}$$
$$- \min_{\tilde{G}_m^{i+1} \supset G_m^i} \left\{ \mathsf{E}\left(T_n(d) \mid \tilde{G}_m^{i+1}\right) \right\}.$$

Put $\hat{G}_m^{i+1} = \arg\max \mathsf{E}(T_n(d) \mid \tilde{G}_m^{i+1})$, $\bar{G}_m^{i+1} = \arg\min \mathsf{E}(T_n(d) \mid \tilde{G}_m^{i+1})$. It is sufficient to estimate the difference $\mathsf{E}(T_n(d) \mid \hat{G}_m^{i+1}) - \mathsf{E}(T_n(d) \mid \bar{G}_m^{i+1})$.

For $i + 1 \leq t \leq n$ put

$$\delta_t^i(d) = \mathsf{E}(T_t(d) \mid \hat{G}_m^{i+1}) - \mathsf{E}(T_t(d) \mid \bar{G}_m^{i+1}).$$

First, let us note that for $n \leq W \cdot d^2$ (the value of constant W will be defined later) we have $\delta_n^i(d) \leq \frac{2mn}{d} \cdot \left(\frac{m(m-1)}{2} + d\,m \right) \leq 4m^2 n \leq Md^2 \leq Md^2 \log(n) \leq Md^2 n^{2A-1}$ (since we have at most $\frac{2mn}{d}$ vertices of degree d, and each vertex of degree d has at most $\frac{m(m-1)}{2}$ triangles when this vertex appears plus at each step we get a triangle only if we hit both the vertex under consideration and a neighbor of this vertex, and our vertex degree is equal to d, therefore we get at most $d\,m$ triangles) for some constant M which depends only on W and m.

It remains to estimate $\delta_n^i(d)$ for $n > W d^2$. Consider the case $2A < 1$. We want to prove that $\delta_n^i(d) \leq Md^2$ for $n > W d^2$ by induction. Suppose that $n = i+1$. Fix G_m^i. Graphs $\hat{G}_m^{i+1}$ and $\bar{G}_m^{i+1}$ are obtained from the graph G_m^i by adding the vertex $i+1$ and m edges. These m edges can affect the number of triangles on at most m previous vertices. For example, they can be drown to at most m vertices of degree d and decrease $T_i(d)$ by at most $\frac{m\,d\,(d-1)}{2}$. Such reasonings finally lead to the estimate $\delta_{i+1}^i(d) \leq Md^2$ for some M.

Now let us use the induction. Consider t: $i + 1 \leq t \leq n - 1$, $t > W d^2$ (note that the smaller values of t were already considered). Using similar reasonings as in the proof of Theorem 4 we get:

$$\delta_{t+1}^i(m) = \delta_t^i(m)\,(1 - p_t(m)) + O\left(\frac{1}{t}\right),$$

$$\delta_{t+1}^i(d) = \delta_t^i(d)\,(1 - p_t(d)) + \delta_t^i(d-1)\,p_t^1(d-1)$$
$$+ (d-1) \cdot \left(\mathsf{E}(N_t(d-1) \mid \hat{G}_m^i) - \mathsf{E}(N_t(d-1) \mid \bar{G}_m^i) \right) \cdot \frac{D}{mt}$$
$$+ O\left(\frac{d \cdot \mathsf{E}S(t, d-1)}{t^2} \right) + O\left(\frac{\mathsf{E}T_t(d) \cdot d^2}{t^2} \right) + O\left(\frac{\mathsf{E}N_t(d) \cdot d^3}{t^2} \right).$$

Note that $E(N_t(d) \mid \hat{G}_m^{i+1}) - E(N_t(d) \mid \bar{G}_m^{i+1}) = O\,(d)$ (see [19]) and $ES(t, d - 1) = O\,(t)$. From this recurrent relations it is easy to obtain by induction that $\delta_n^i(d) \le Md^2$ for some M. Indeed,

$$\delta_{t+1}^i(m) \le Mm^2\,(1 - p_t(m)) + \frac{C_1}{t} \le Mm^2 \left(1 - \frac{Am + B}{t} + \frac{C_2}{t^2}\right) + \frac{C_1}{t} \le Mm^2$$

for sufficiently large M. By C_i, $i = 1, 2, \ldots$, we denote some positive constants. For $d > m$ we get

$$\delta_{t+1}^i(d) \le Md^2(1 - p_t(d)) + M(d-1)^2 p_t^1(d-1) + C_3\frac{d^2}{t} + C_4\frac{d^4}{t^2}$$

$$\le Md^2\left(1 - \frac{Ad + B}{t} + C_5\frac{d^2}{t^2}\right) + M(d-1)^2\left(\frac{A(d-1) + B}{t} + C_6\frac{d^2}{t^2}\right) + C_3\frac{d^2}{t}$$

$$+ C_4\frac{d^4}{t^2} \le Md^2 + \frac{M}{t}\left(A(-3d^2 + 3d - 1) + B(-2d + 1) + C_7\frac{d^4}{t} + C_3\frac{d^2}{M} + C_4\frac{d^4}{Mt}\right)$$

$$\le Md^2 + \frac{M}{t}\left(\left(-3A + C_7\frac{d^2}{t} + \frac{C_3}{M} + C_4\frac{d^2}{Mt}\right) \cdot d^2\right)$$

$$+ (3A - 2B) \cdot d + (B - A)) \le Md^2.$$

for sufficiently large W and M.

In the case $2A = 1$ we have $ES(t, d-1) = O\,(t\log(t))$ and we get the following inequalities:

$$\delta_{t+1}^i(m) \le Mm^2\log(t)\,(1 - p_t(m)) + \frac{C_1\log(t)}{t} \le Mm^2\log(t + 1),$$

$$\delta_{t+1}^i(d) \le Md^2\log(t)(1 - p_t(d)) + M(d-1)^2\log(t)\,p_t^1(d-1)$$

$$+ C_2\frac{d^2}{t} + C_3\frac{d\log(t)}{t} + C_4\frac{d^4\log(t)}{t^2} \le Md^2\log(t + 1).$$

In the case $2A > 1$ we have $ES(t, d - 1) = O\,(t^{2A})$ and we get the following inequalities:

$$\delta_{t+1}^i(m) \le Mm^2 t^{2A-1}\,(1 - p_t(m)) + \frac{C_1 t^{2A-1}}{t} \le Mm^2(t + 1)^{2A-1},$$

$$\delta_{t+1}^i(d) \le Md^2 t^{2A-1}(1 - p_t(d)) + M(d-1)^2\,t^{2A-1} p_t^1(d-1)$$

$$+ C_2\frac{d^2}{t} + C_3\frac{d \cdot t^{2A-1}}{t} + C_4\frac{d^4 t^{2A-1}}{t^2} \le Md^2(t + 1)^{2A-1}.$$

This concludes the proof of Theorem 5.

5 Conclusion

In this paper, we study the local clustering coefficient $C(d)$ for the vertices of degree d in the T-subclass of the PA-class of models. Despite the fact that the T-subclass generalizes many different models, we are able to analyze the local clustering coefficient for all these models. Namely, we proved that $C(d)$ asymptotically decreases as $\frac{2D}{Am} \cdot d^{-1}$. In particular, this result implies that one cannot change the exponent -1 by varying the parameters A, D, and m. This basically means that preferential attachment models in general are not flexible enough to model $C(d) \sim d^{-\psi}$ with $\psi \neq 1$.

We would also like to mention the connection between the obtained result and the notion of *weak* and *strong transitivity* introduced in [21]. It was shown in [22] that percolation properties of a network are defined by the type (weak or strong) of its connectivity. Interestingly, a model from the T-subclass can belong to either weak or strong transitivity class: if $2D < Am$, then we obtain the weak transitivity; if $2D > Am$, then we obtain the strong transitivity.

References

1. Albert, R., Barabási, A.-L.: Statistical mechanics of complex networks. Rev. Mod. Phys. **74**, 47–97 (2002)
2. Bansal, S., Khandelwal, S., Meyers, L.A.: Exploring biological network structure with clustered random networks. BMC Bioinf. **10**, 405 (2009)
3. Barabási, A.-L., Albert, R.: Emergence of scaling in random networks. Sci. **286**(5439), 509–512 (1999)
4. Barabási, A.-L., Albert, R., Jeong, H.: Mean-field theory for scale-free random networks. Phys. A **272**(1–2), 173–187 (1999)
5. Albert, R., Jeong, H., Barabási, A.-L.: Internet: diameter of the world-wide web. Nat. **401**, 130–131 (1999)
6. Boccaletti, S., Latora, V., Moreno, Y., Chavez, M., Hwang, D.-U.: Complex networks: structure and dynamics. Phys. Rep. **424**(45), 175–308 (2006)
7. Bollobás, B., Riordan, O.M.: Mathematical results on scale-free random graphs. In: Handbook of Graphs and Networks: From the Genome to the Internet (2003)
8. Bollobás, B., Riordan, O.M., Spencer, J., Tusnády, G.: The degree sequence of a scale-free random graph process. Random Struct. Algorithms **18**(3), 279–290 (2001)
9. Borgs, C., Brautbar, M., Chayes, J., Khanna, S., Lucier, B.: The power of local information in social networks. Preprint (2012)
10. Broder, A., Kumar, R., Maghoul, F., Raghavan, P., Rajagopalan, S., Stata, R., Tomkins, A., Wiener, J.: Graph structure in the web. Comput. Netw. **33**(16), 309–320 (2000)
11. Buckley, P.G., Osthus, D.: Popularity based random graph models leading to a scale-free degree sequence. Discrete Math. **282**, 53–63 (2004)
12. Catanzaro, M., Caldarelli, G., Pietronero, L.: Assortative model for social networks. Phys. Rev. E **70**, 037101 (2004)
13. Leskovec, J.: Dynamics of large networks. ProQuest (2008)
14. Faloutsos, M., Faloutsos, P., Faloutsos, C.: On power-law relationships of the internet topology. In: Proceedings of SIGCOMM (1999)

15. Girvan, M., Newman, M.E.: Community structure in social and biological networks. Proc. Nat. Acad. Sci. **99**(12), 7821–7826 (2002)
16. Holme, P., Kim, B.J.: Growing scale-free networks with tunable clustering. Phys. Rev. E **65**(2), 026107 (2002)
17. Newman, M.E.J.: Pareto distributions and Zipf's law. Contemp. Phys. **46**(5), 323–351 (2005)
18. Newman, M.E.J.: The structure and function of complex networks. SIAM Rev. **45**(2), 167–256 (2003)
19. Ostroumova, L., Ryabchenko, A., Samosvat, E.: Generalized preferential attachment: tunable power-law degree distribution and clustering coefficient. In: Bonato, A., Mitzenmacher, M., Prałat, P. (eds.) WAW 2013. LNCS, vol. 8305, pp. 185–202. Springer, Heidelberg (2013)
20. Ravasz, E., Barabási, A.-L.: Hierarchical organization in complex networks. Phys. Rev. E **67**(2), 26112 (2003)
21. Serrano, M.A., Boguñá, M.: Clustering in complex networks. I. General formalism. Phys. Rev. E **74**, 056114 (2006)
22. Serrano, M.A., Boguñá, M.: Clustering in complex networks. II. Percolation properties. Phys. Rev. E **74**, 056115 (2006)
23. Vázquez, A., Pastor-Satorras, R., Vespignani, A.: Large-scale topological and dynamical properties of the internet. Phys. Rev. E **65**, 066130 (2002)
24. Watts, D.J., Strogatz, S.H.: Collective dynamics of 'small-world' networks. Nature **393**, 440–442 (1998)
25. Zhou, T., Yan, G., Wang, B.-H.: Maximal planar networks with large clustering coefficient and power-law degree distribution. Phys. Rev. E **71**(4), 046141 (2005)

Hyperbolicity, Degeneracy, and Expansion of Random Intersection Graphs

Matthew Farrell[1], Timothy D. Goodrich[2], Nathan Lemons[3], Felix Reidl[4], Fernando Sánchez Villaamil[4], and Blair D. Sullivan[2(✉)]

[1] Department of Applied Mathematics, University of Washington, Seattle, WA, USA
msf9@uw.edu
[2] Department of Computer Science, North Carolina State University, Raleigh, NC, USA
{tdgoodri,blair_sullivan}@ncsu.edu
[3] Theoretical Division, Los Alamos National Laboratory, Los Alamos, NM, USA
nlemons@gmail.com
[4] Theoretical Computer Science, RWTH Aachen, Aachen, Germany
{reidl,fernando.sanchez}@cs.rwth-aachen.de

Abstract. We establish the conditions under which several algorithmically exploitable structural features hold for random intersection graphs, a natural model for many real-world networks where edges correspond to shared attributes. Specifically, we fully characterize the degeneracy of random intersection graphs, and prove that the model asymptotically almost surely produces graphs with hyperbolicity at least $\log n$. Further, we prove that when degenerate, the graphs generated by this model belong to a bounded-expansion graph class with high probability, a property particularly suitable for the design of linear time algorithms.

1 Introduction

There has been a recent surge of interest in analyzing large graphs, stemming from the rise in popularity (and scale) of social networks and significant growth of relational data in science and engineering fields (e.g. gene expressions, cyber-security logs, and neural connectomes). One significant challenge in the field is the lack of deep understanding of the underlying structure of various classes of real-world networks. Here, we focus on two structural characteristics that can be exploited algorithmically: *bounded expansion* and *hyperbolicity*.

A graph class has *bounded expansion*[1] if for every member G, one cannot form arbitrarily dense graphs by contracting subgraphs of small radius. Intuitively, this naturally corresponds to sparse interactions between locally dense clusters or communities. Formally, the density of every minor of G is bounded by a function of the *depth* of that minor (the maximum radius of its branch sets). Bounded expansion offers a structural generalization of both bounded-degree and graphs excluding a (topological) minor. Algorithmically, this property is

---

[1] Not related to the notion of expander graphs.

© Springer International Publishing Switzerland 2015
D.F. Gleich et al. (Eds.): WAW 2015, LNCS 9479, pp. 29–41, 2015.
DOI: 10.1007/978-3-319-26784-5_3

extremely useful: every first-order-definable problem is decidable in linear fpt-time in these classes [10]. For example, counting the number of appearances of a fixed pattern graph as a subgraph can be computed in linear time [9,22]. We also consider δ-*hyperbolicity*, which restricts the structure of shortest-path distances in the graph to be tree-like. Hyperbolicity is closely tied to treelength [7], but unrelated to measures of structural density such as bounded expansion. Algorithms for graph classes of bounded hyperbolicity often exploit computable approximate distance trees [7] or greedy routing [17]. Both of these properties present challenges for empirical evaluation—bounded expansion is only defined with respect to graph classes (not for single instances), and hyperbolicity is an extremal statistic whose $O(n^4)$ computation is infeasible for many of today's large data sets. As is typical in the study of network structure, we instead ask how the properties behave with respect to randomized models which are designed to mimic aspects of network formation and structure.

In this paper, we consider the *random intersection graph model* introduced by Karoński, Scheinerman, and Singer-Cohen [16,27] which has recently attracted significant attention in the literature [4,8,12,14,26]. *Random intersection graphs* are based on the premise that network edges often represent underlying shared interests or attributes. The model first creates a bipartite object-attribute graph $B = (V, A, E)$ by adding edges uniformly at random with a fixed per-edge probability $p(\alpha)$, then considers the *intersection graph*: $G := (V, E')$ where $xy \in E'$ iff the neighborhoods of the vertices x, y in B have a non-empty intersection. The parameter α controls both the ratio of attributes to objects and the probability p: for n objects the number of attributes m is proportional to n^α and the probability p to $n^{-(1+\alpha)/2}$.

Recently, random intersection graphs have gained popularity in modeling real-world data. For example, Zhao et al. use random intersection graphs to model the Eschenauer-Gligor (EG) key predistribution scheme for secure connections in wireless networking [29,31,32]. Given sensor nodes and limitations on their communication ranges (visibility), the problem is to design a topology for optimal communication. By modeling sensor link unreliability and transmission constraints with Erdős-Rényi and random geometric graphs, respectively, the topology of the final EG scheme can be computed with a random intersection graph on the sensors. Random intersection graphs have also been used in modeling cybersecurity [2], the spread of epidemics [1], social networks [24,28], and clustering [30].

Beyond inherently modeling native structure in many real-world applications, random intersection graphs also have (1) relative mathematical tractability yielded by the independence of events in the underlying edge creation process, and (2) the ability to generate graphs with key structural properties matching real data—namely sparsity, (tunable) clustering and assortativity [3,4,8]. Together, these features allow the design of random graph generators that produce graphs with specific and well-understood properties such as connectivity and degree distribution. Extending this mathematical understanding, we examine more complex structural properties such as expansion and hyperbolicity. Specifically, in this paper we present the following results on the structure of random intersection graphs:

(i) For $\alpha \leq 1$, with high probability (w.h.p.), random intersection graphs are somewhere dense (and thus do not have bounded expansion) and have unbounded degeneracy.

(ii) For $\alpha > 1$, w.h.p. random intersection graphs have bounded expansion (and thus constant degeneracy).

(iii) Under reasonable restrictions on the constants in the model, random intersection graphs have hyperbolicity $\Omega(\log n)$ asymptotically almost surely.

In particular, the second result strengthens the original claim that the model generates sparse graphs for $\alpha > 1$, by establishing they are in fact *structurally sparse* in a robust sense. We note that random intersection graphs only exhibit tunable clustering when $\alpha = 1$ [8], so our results strongly support the following: Homogeneous random intersection graphs fail at being sparse and having tunable clustering simultaneously[2]. Finally, we note that the third result is negative—our bound implies a $\log n$ lower bound on the treelength [7].

2   Preliminaries

We start with a few necessary definitions and lemmas, covering each of the key ideas in the paper (random intersection graphs, degeneracy, expansion, and hyperbolicity).

Since this paper is concerned with asymptotic results, we use standard asymptotic terminology: for each integer n, let $\mathcal{G}_n$ define a distribution on graphs with n vertices (for example, coming from a random graph model). We say the events E_n defined on $\mathcal{G}_n$ hold *asymptotically almost surely* (a.a.s.) if $\lim_{n\to\infty} \mathbb{P}[E_n] = 1$. Furthermore, we say an event occurs *with high probability* (w.h.p.) if for any $c \geq 1$ the event occurs with probability at least $1 - O(n^{-c})$. As a shorthand, we will simply say that $\mathcal{G}_n$ *has some property* a.a.s.(or w.h.p.).

2.1   Random Intersection Graphs

A wide variety of random intersection graph models have been defined in the literature; in this paper, we restrict our attention to the most well-studied of these, $G(n, m, p)$, which is defined as follows:

Definition 1 (Random Intersection Graph Model). *Fix positive constants* α, β *and* γ. *Let* B *be a random bipartite graph on parts of size* n *and* $m = \beta n^\alpha$ *with each edge present independently with probability* $p = \gamma n^{-(1+\alpha)/2}$. *Let* V *(the nodes) denote the part of size* n *and* A *(the attributes) the part of size* m. *The associated* random intersection graph $G = G(n, m, p)$ *is defined on the nodes* V: *two nodes are connected in* G *if they share (are both adjacent to in* B*) at least one attribute in* A.

---

[2] It should be noted that constant clustering and bounded expansion are not orthogonal [9].

We note that $G(n, m, p)$ defines a distribution $\mathcal{G}_n$ on graphs with n vertices. The notation $G = G(n, m, p)$ denotes a graph G that is randomly sampled from the distribution $\mathcal{G}_n$. Throughout the manuscript, given a random intersection graph $G(n, m, p)$ we will refer to B, the associated bipartite graph on n nodes and m attributes from which G is formed.

In order to work with graph classes formed by the random intersection graph model, we will need a technical result that bounds the number of attributes in the neighborhood of a subset of nodes around its expected value. These lemmas and their proofs can be found in the full version of this paper [11].

2.2   Degeneracy and Expansion

Although it is widely accepted that complex networks tend to be sparse (in terms of edge density), this property does not suffice for reducing the algorithmic complexity of NP-hard analysis tasks. In order to take advantage of parameterized algorithms, we focus on *structural* sparseness. For instance, it is not enough for a graph to be sparse on average, we would also like it to have sparse subgraphs. This motivates a very general class of structurally sparse graphs—those of bounded *degeneracy*: A graph is d-degenerate if every subgraph has a vertex of degree at most d. It is easy to see that the degeneracy is lower-bounded by the size of the largest clique. Thus, the degeneracy of intersection graphs is bounded below by the maximum attribute degree in the associated bipartite graph (since each attribute contributes a complete subgraph of size equal to its degree to the intersection graph). For certain parameter values, this lower bound will, w.h.p., give the correct order of magnitude of the degeneracy of the graph.

Bounded degeneracy however is often too weak a structural guarantee for the design fast algorithms. Here we focus on the stronger structural property of *bounded expansion* which provides a rich framework of algorithmic tools [19]. In the context of networks, bounded expansion captures the idea that networks decompose into small dense structures (e.g. communities) connected by a sparse global structure. More formally, we characterize bounded-expansion classes using special graph minors and an associated density measure (the grad).

Definition 2 (Shallow topological minor, nails, subdivision vertices). *A graph M is an r-shallow topological minor of G if a $(\leq 2r)$-subdivision of M is isomorphic to a subgraph H' of G. We call H' a model of M in G. For simplicity, we assume by default that $V(M) \subseteq V(H')$ such that the isomorphism between M and H' is the identity when restricted to $V(M)$. The vertices $V(M)$ are called* nails *and the vertices $V(H') \setminus V(M)$* subdivision vertices. *The set of all r-shallow topological minors of a graph G is denoted by $G \widetilde{\triangledown} r$.*

Definition 3 (Topological grad). *For a graph G and integer $r \geq 0$, the topological greatest reduced average density (grad) at depth r is defined as $\widetilde{\nabla}_r(G) = \max_{H \in G \widetilde{\triangledown} r} |E(H)|/|V(H)|$. For a graph class $\mathcal{G}$, define $\widetilde{\nabla}_r(\mathcal{G}) = \sup_{G \in \mathcal{G}} \widetilde{\nabla}_r(G)$.*

Definition 4 (Bounded expansion). *A graph class $\mathcal{G}$ has bounded expansion if there exists a function f such that for all r, we have $\widetilde{\nabla}_r(\mathcal{G}) < f(r)$.*

When introduced, bounded expansion was originally defined using an equivalent characterization based on the notion of *shallow minors* (cf. [19]): H is a r-shallow minor of G if H can be obtained from G by contracting disjoint subgraphs of radius at most r. In the context of our paper, however, the topological shallow minor variant proves more useful, and we restrict our attention to this setting. Let us point out that bounded expansion implies bounded degeneracy, with $2f(0)$ being an upper bound on the degeneracy of the graphs. In particular, $G \widetilde{\triangledown} 0$ is the set of all subgraphs of G.

Finally, in order to characterize when the model is not structurally sparse, we define another class in the hierarchy – *nowhere dense* is a generalization of bounded expansion in which we measure the *clique number* instead of the edge density of shallow minors. Let $\omega(G)$ denote the size of the largest complete subgraph of a graph G and let $\omega(\mathcal{G}) = \sup_{G \in \mathcal{G}} \omega(G)$ be the natural extension to graph classes $\mathcal{G}$.

Definition 5 (Nowhere dense [20,21]). *A graph class $\mathcal{G}$ is nowhere dense if there exists a function f such that for all $r \in \mathbf{N}$ it holds that $\omega(\mathcal{G} \widetilde{\triangledown} r) < f(r)$.*

A graph class is *somewhere dense* precisely when it is not nowhere dense. While in general a graph class with unbounded degeneracy is not necessarily somewhere dense, the negative proofs presented here show that members of the graph class contain large cliques w.h.p. This simultaneously implies unbounded degeneracy and that the class is somewhere dense (as a clique is a 0-subdivision of itself). Consequently, we prove a clear dichotomy: random intersection graphs are either structurally sparse or somewhere dense.

2.3   Gromov's Hyperbolicity

The concept of δ-hyperbolicity was introduced by Gromov in the context of geometric group theory [13]. It captures how "tree-like" a graph is in terms of its metric structure, and has received attention in the analysis of networks and informatics graphs. We refer the reader to [6,15,17,18], and references therein, for details on the motivating network applications.

There are several ways of characterizing δ-hyperbolic metric spaces, all of which are equivalent up to constant factors [5,7,13]. Since graphs are naturally geodesic metric spaces when distance is defined using shortest paths, we will use the definition based on δ-slim triangles (originally attributed to Rips [5,13]).

Definition 6. *A graph $G = (V, E)$ is δ-hyperbolic if for all $x, y, z \in V$, for every choice of geodesic (shortest) paths between them —denoted $P[x, y], P[x, z],$ $P[y, z]$ —we have $\forall v \in P[x, y], \exists w \in P[x, z] \cup P[z, y] : d_G(v, w) \leq \delta$, where $d_G(u, v)$ is shortest-path distance in G.*

That is, if G is δ-hyperbolic, then for each triple of vertices x, y, z, and every choice of three shortest paths connecting them pairwise, each point on the shortest path from x to y must be within distance δ of a point on one of the other paths. The *hyperbolicity* of a graph G is the minimum $\delta \geq 0$ so that G is δ-hyperbolic. Note that a trivial upper bound on the hyperbolicity is half the diameter (this is true for any graph).

In this paper we give lower bounds for the hyperbolicity of the graphs in $G(n, m, p)$. We believe (but do not prove) these bounds are asymptotically the correct order of magnitude (e.g. also upper bounds). This would require that the diameter of connected components is also logarithmic in n, which has been shown for a similar model [25].

3 Structural Sparsity of Random Intersection Graphs

In this section we will characterize a clear break in the sparsity of graphs generated by $G(n, m, p)$, depending on the relative grown rates of the nodes and attributes in B). In each case, we analyze (probabilistically) the degeneracy and expansion of the generated class.

Theorem 1. *Fix constants α, β and γ. Let $m = \beta n^\alpha$ and $p = \gamma n^{-(1+\alpha)/2}$. Let $G = G(n, m, p)$. Then the following hold w.h.p.*

(i) *If $\alpha < 1$, $G(n, m, p)$ is somewhere dense and G has degeneracy $\Omega(\gamma n^{(1-\alpha)/2})$.*
(ii) *If $\alpha = 1$, $G(n, m, p)$ is somewhere dense and G has degeneracy $\Omega(\frac{\log n}{\log \log n})$.*
(iii) *If $\alpha > 1$, $G(n, m, p)$ has bounded expansion and thus G has degeneracy $O(1)$.*

We prove Theorem 1 separately for each of the three ranges of α. When $\alpha \leq 1$, we prove that w.h.p. the random intersection graph model generates graph classes with unbounded degeneracy by establishing the existence of a high-degree attribute in the associated bipartite graph (thus lower-bounding the clique number). These proofs can be found in the full version of this paper [11].

In the remainder of this section, we focus on the case when $\alpha > 1$, as this is the parameter range in which the model generates sparse graphs. Here, we present the general structure of the proof of Theorem 1 for the case $\alpha > 1$. Detailed proofs for all lemmas and theorems can be found in the full version of this paper [11].

Before beginning, we note that if $G(n, m, p)$ has bounded expansion w.h.p., then for any $p' \leq p$ and $m' \leq m$ it follows that w.h.p. $G(n, m', p')$ also has bounded expansion by a simple coupling argument. Thus we can assume without loss of generality that both γ and β are greater than one. For the remainder of this section, we fix the parameters $\gamma, \beta, \alpha > 1$, the resulting number of attributes $m = \beta n^\alpha$ and the per-edge probability $p = \gamma n^{-(1+\alpha)/2}$.

3.1 Bounded Attribute-Degrees

As mentioned before, for a random intersection graph to be degenerate, the attributes of the associated bipartite graph must have bounded degree. We prove that w.h.p., this necessary condition is satisfied.

Lemma 2. *Let* $c \geq 1$ *be a constant such that* $2\frac{\alpha+c}{\alpha-1} > \beta\gamma e$. *Then the probability that there exists an attribute in the bipartite graph associated with* $G(n, m, p)$ *of degree higher than* $2\frac{\alpha+c}{\alpha-1}$ *is* $O(n^{-c})$.

This allows us to assume for the remainder of the proof that the maximum attribute degree is bounded.

3.2 Alternative Characterization of Bounded Expansion

We now state a characterization of bounded expansion which is often helpful in establishing the property for classes formed by random graph models.

Proposition 1 [22,23]. *A class* $\mathcal{C}$ *of graphs has bounded expansion if and only if there exists real-valued functions* $f_1, f_2, f_3, f_4 \colon \mathbf{R} \to \mathbf{R}^+$ *such that the following two conditions hold:*

(i) For all positive ϵ *and for all graphs* $G \in \mathcal{C}$ *with* $|V(G)| > f_1(\epsilon)$, *it holds that*
$$\frac{1}{|V(G)|} \cdot |\{v \in V(G) \colon \deg(v) \geq f_2(\epsilon)\}| \leq \epsilon.$$
(ii) For all $r \in \mathbf{N}$ *and for all* $H \subseteq G \in \mathcal{C}$ *with* $\widetilde{\nabla}_r(H) > f_3(r)$, *it follows that* $|V(H)| \geq f_4(r) \cdot |V(G)|$.

Intuitively, this states that any class of graphs with bounded expansion is characterized by two properties:

(i) all sufficiently large members of the class have a small fraction of vertices of large degree;
(ii) all subgraphs of $G \in \mathcal{C}$ whose shallow topological minors are sufficiently dense must necessarily span a large fraction of the vertices of G.

3.3 Stable r-Subdivisions

In order to disprove the existence of an r-shallow topological minor of a certain density δ, we introduce a stronger topological structure.

Definition 7 (Stable r-subdivision). *Given graphs* G, H *we say that* G *contains* H *as a stable* r-*subdivision if* G *contains* H *as a* $\frac{r}{2}$-*shallow topological minor with model* G' *such that every path in* G' *corresponding to an edge in* H *has exactly length* $r + 1$ *and is an induced path in* G.

A stable r-subdivision is by definition a shallow topological minor, thus the existence of an r-subdivision of density δ implies that $\widetilde{\nabla}_{\frac{r}{2}}(G) \geq \delta$. We prove that the densities are also related in the other direction.

Lemma 3. *A graph* G *with* $\widetilde{\nabla}_{\frac{r}{2}}(G) \geq \delta$ *contains a stable* i-*subdivision of density at least* $\delta/(r+1)$ *for some* $i \in \{0, \dots, r\}$.

To show that a graph has no r-shallow minor of density δ, it now suffices to prove that no stable i-subdivision of density $\delta/(2r+1)$ exists for any $i \in \{0, \ldots, 2r\}$. We note that the other direction would not work, since the existence of a stable i-subdivision for some $i \in \{0, \ldots, 2r\}$ of density $\delta/(2r+1)$ does not imply the existence of an r-shallow topological minor of density δ.

We now establish the probability of having this structure in the random intersection graph model, noting that the following structural result is surprisingly useful, and appears to have promising applications beyond this work.

Theorem 4. *Let $c \geq 1$ be a constant, $\Delta := 2\frac{\alpha+c}{\alpha-1}$, and $\phi = (6e\Delta\beta\gamma r\delta)^{5r\delta\frac{2}{(\alpha-1)}}$. Assuming that no attribute has degree greater than Δ, the probability that $G(n, m, p)$ contains a stable r-subdivision with k nails for $r \geq 1$ and of density $\delta > 1$ is at most $r\delta k \cdot (\phi/n)^{\frac{\alpha-1}{2}k}$.*

Proof sketch. We argue that a dense subdivision in G implies the existence of a dense subgraph in the associated bipartite graph. We show this by considering the existence of a stable r-subdivision where all paths are induced, which is generated by a minimal number of attributes. Notice that if a model of some graph H exists, so does a model with these properties. This allows us to only consider attributes with minimum degree two, since every edge in the path is generated by a different attribute. This is key to giving an upper bound for the probability of the existence of such a dense subgraph in the bipartite graph and prove the theorem.

3.4 Density

Before turning to our main result, we need two more lemmas that establish the probability of graphs generated using $G(n, m, p)$ have special types of dense subgraphs.

Lemma 5. *Let $c \geq 1$ be a constant and let $\Delta := 2\frac{\alpha+c}{\alpha-1}$. For $u \leq m, k \leq n$, the probability that the bipartite graph associated with $G(n, m, p)$ contains u attributes of degree $\leq \Delta$ that generate at least $\rho \geq u$ edges between k fixed vertices is at most*

$$\left(\frac{e^{\Delta+1}\gamma^\Delta \Delta\beta}{u/k}\right)^u \left(\frac{k}{n}\right)^u.$$

We note that it is perhaps surprising that ρ disappears in the upper bound given above. Since we are assuming that the degree of the attributes is bounded by Δ, the number of attributes u must be at least $\rho/\binom{\Delta}{2}$. Thus the ρ reappears upon expansion. Since we can bound the degree of the attributes w.h.p. when $\alpha > 1$ this theorem is generally applicable to sparse random intersection graphs.

The following Lemma is a rather straightforward consequence of Lemma 5.

Lemma 6. *Let $c \geq 1$ be a constant, $\Delta := 2\frac{\alpha+c}{\alpha-1}$ and $\delta > e^{\Delta+1}\gamma^\Delta \Delta^3\beta$. Then the probability that $G(n, m, p)$ contains a subgraph of density δ on k nodes is at most $\delta k(k/n)^{\delta k/\Delta^2}$.*

3.5   Main Result

Theorem 7. *Fix positive constants $\alpha > 1$, β and γ. Then w.h.p. the class of random intersection graphs $G(n, m, p)$ defined by these constants has bounded expansion.*

Proof sketch. We show the two conditions of Proposition 1 are satisfied in Lemmas 8 and 9, respectively. The proof of Lemma 8 can be found in the full version of this paper [11].

Lemma 8. *Let $c \geq 1$ be a constant, $\Delta := 2\frac{\alpha+c}{\alpha-1}$ and λ be a constant bigger than $\max\{2e^{\Delta+2}\gamma^{\Delta}\Delta\beta, c\}$. For $G = \mathcal{G}(n, m, p)$ and for all $\epsilon > 0$ it holds with probability $O(n^{-c})$ that $|V(G)|^{-1} \cdot |\{v \in V(G): \deg(v) \geq \frac{2\lambda\Delta^2}{\epsilon}\}| \leq \epsilon$.*

Lemma 9. *Let $c \geq 1$ be a constant, $\Delta := 2\frac{\alpha+c}{\alpha-1}$, $\Delta^2 := \binom{\Delta}{2}$, ϕ be defined as in Theorem 4 and $\delta_r > (2r+1) \cdot \max\{e^{\Delta+1}\gamma^{\Delta}\Delta\Delta^2\beta, (c+1)\Delta^2\}$. Then for every $r \in \mathbf{N}^+$, for every $0 < \epsilon < e^{-3}$, and for every $H \subseteq G = G(n, m, p)$ with $|H| < \epsilon n$ it holds with probability $O(n^{-c})$ that $\widetilde{\nabla}_r(H) \geq \delta_r$.*

Proof. By Lemma 2 we can disregard any graph whose associated bipartite graph has an attribute of degree greater than Δ. By Lemma 3 if G contains an r-shallow topological minor of density δ_r then for some $i \in \{0, \ldots, 2r\}$ there exists a stable i-subdivision of density $\delta_r/(2r+1)$. We can then bound the probability of a r-shallow topological minor by bounding the probability of a stable i-subdivision of density $\delta_r/(2r+1)$.

From Lemma 6 we know that the probability of a 0-shallow topological minor on k nails is bounded by

$$\binom{n}{k}\delta_r k \left(\frac{k}{n}\right)^{\frac{\delta_r k}{\Delta^2}}. \tag{1}$$

By Theorem 4, the probability for an i-subdivision of density $\delta_r/(2r+1)$ for $i \in \{1, \ldots, 2r\}$ is bounded by $r\delta_r k \cdot (\phi/n)^{k(\alpha-1)/2}$. Taking the union bound of these two events gives us a total bound of

$$\binom{n}{k}\delta_r k \left(\frac{k}{n}\right)^{\frac{\delta_r k}{\Delta^2}} + (2r+1)r\delta_r k \cdot \left(\frac{\phi}{n}\right)^{\frac{\alpha-1}{2}k} \tag{2}$$

for the probability of a dense subgraph or subdivision on k vertices to appear. Taking the union bound over all k we obtain for the first summand that

$$\sum_{k=1}^{\epsilon n} \binom{n}{k}\delta_r k \left(\frac{k}{n}\right)^{\frac{\delta_r k}{\Delta^2}} \leq \delta_r \sum_{k=1}^{\epsilon n} \frac{n^k e^k}{k^k} \frac{k^{(c+1)k+1}}{n^{(c+1)k}}. \tag{3}$$

Since δ_r is a constant, it suffices that the sum in (3) is in $O(n^{-c})$. We will show this is bounded by a geometric sum by considering the ratio of two consecutive summands:

$$\frac{e^{k+1}(k+1)^{c(k+1)+1}}{n^{c(k+1)}} \frac{n^{ck}}{e^k k^{ck+1}} = e\frac{(k(1+1/k))^{ck+c+1}}{n^c k^{ck+1}} \leq e^{2c+1}\frac{k^c}{n^c} \leq e^{2c+1}\epsilon^c. \tag{4}$$

Here we used the fact that $(1 + 1/k)^{ck+c+1} \leq e^c(1 + 1/k)^{c+1}$ and that $(1 + 1/k)^{c+1} \leq e^c$ for $k \geq 2$ and $c \geq 1$.

Since this is smaller than one when $\epsilon < e^{-2}$ and $c \geq 1$, the summands decrease geometrically. Hence its largest element (i.e. the summand for $k = 1$) dominates the total value of the sum, more precisely, there exists a constant ξ (depending on α and c) such that

$$\sum_{k=1}^{\epsilon n} \frac{e^k k^{ck+1}}{n^{ck}} \leq \xi \frac{e}{n^c} = O(n^{-c}). \tag{5}$$

We now turn to the second summand. It is easy to see by the same methods as before that this sum is also geometric for $n > \phi^{(\alpha+1)/2}$ and as such there exists a constant ξ' which multiplied with the first element bounds the sum. An r-shallow topological minor of density δ_r has at least $2\delta_r$ nails, thus we can assume $k \geq 2\delta_r$. Since $\delta_r > (c+1)\Delta^2 \geq c/(\alpha - 1)$, we have:

$$\sum_{k=2\delta_r}^{\epsilon n} (2r+1)r\delta_r k \cdot \left(\frac{\phi}{n}\right)^{\frac{\alpha-1}{2}k} \leq \frac{\xi'(2r+1)\phi^{\delta_r}}{n^{(\alpha-1)\delta_r}} \leq \frac{\xi'(2r+1)\phi^{\delta_r}}{n^c} = O(n^{-c}). \tag{6}$$

Combining (5) and (6), Eq. (2) is bounded by $O(n^{-c})$, as claimed. $\qquad\square$

4   Hyperbolicity

We now turn to the question of whether the structure of the shortest-path distances in random intersection graphs is tree-like (using Gromov's δ-hyperbolicity as defined in Sect. 2.3) where we establish a negative result by giving a logarithmic lower bound, for all values of α. The details of the proof can be found in the full version of the paper [11]. Our approach is based on a special type of path, which gives natural lower bounds on the hyperbolicity.

Definition 8. *Let $G = G(n, m, p)$ be a random intersection graph. The k-path $P = v_1, v_2, \ldots, v_{k+1}$ in G is called a k-special path if all the internal vertices of P have degree two in G and there exists another disjoint path connecting v_1 and v_{k+1} in G. We allow for the second path to have length 0: this occurs if P is a k-cycle such that all but one vertex of P has degree two in G.*

Lemma 10. *Let k be a positive integer and $G = G(n, m, p)$. If G contains a k-special path, then G has hyperbolicity at least $\lfloor \frac{k}{4} \rfloor$.*

Proof. Let P be a k-special path in G. By definition, P is part of some cycle C of length at least k. Without loss of generality, we can suppose that the length of C is exactly k. Then, $v = v_1$, satisfies

$$\forall u \in P_G[v_{\lfloor k/4 \rfloor}, v_{\lceil k/2 \rceil}] \cup P_G[v_{\lceil k/2 \rceil}, v_{\lceil 3k/4 \rceil}], \ |u - v|_G \geq \lfloor k/4 \rfloor. \tag{7}$$

As $v_1 \in P[v_{\lfloor k/4 \rfloor}, v_{\lceil 3k/4 \rceil}]$, this shows the hyperbolicity of G is at least $\lfloor k/4 \rfloor$. $\square$

Showing that these structures exist in an intersection graph is non-trivial, but crucial for our proof of the following theorem.

Theorem 11. *Fix constants α, β and γ such that $\gamma^2 \beta > 1$. There exists a constant $\xi > 0$ such that a.a.s., the random intersection graph $G = G(n, m, p)$ with $m = \beta n^\alpha$ and $p = \gamma n^{-(1+\alpha/2)}$ has hyperbolicity*

(i) at least $\xi \log n$ when $\alpha \geq 1$,
(ii) $(1 \pm o(1))\xi \log n$ otherwise.

Proof sketch. We define a variant of the objects from Definition 8—*k-special bipartite paths*—whose existence in the associated bipartite graph implies a lower bound on hyperbolicity in an intersection graph. We show that for a constant $\xi > 0$, when $k = \xi \log n$, a.a.s. there is at least one k-special path in G. In particular, we let S_k denote the number of k-special bipartite paths, and condition on the fact that the exposed graph has a giant component of size at least δn. We prove that $\mathbb{E}[S_{\xi \log n}] = \omega(1)$, and that S_k is tightly concentrated about its mean, using second moment methods.

5   Conclusion and Open Problems

In this paper we have determined the conditions under which random intersection graphs exhibit two types of algorithmically useful structure. We proved that graphs in $G(n, m, p)$ are structurally sparse (have bounded expansion) precisely when the number of attributes in the associated bipartite graph grows faster than the number of nodes ($\alpha > 1$). Moreover, we showed that when the generated graphs are not structurally sparse, they fail to achieve even much weaker notions of sparsity (in fact, w.h.p. they contain large cliques). We conclude that the tool kit stemming from the bounded expansion framework is applicable to sparse real-world networks whose structure is well-modeled by random intersection graphs with $\alpha > 1$.

On the other hand, we showed that the metric structure of random intersection graphs is not tree-like for all values of α: the hyperbolicity (and treelength) grows at least logarithmically in n. While we only determine a lower bound for the hyperbolicity, we believe this to be the correct order of magnitude, as the diameter (a natural upper bound for the hyperbolicity) of similar model of random intersection graphs was shown to be $O(\log n)$ [25] for a similar range of parameter values.

A question that naturally arises from these results is if structural sparsity should be an expected characteristic of practically relevant random graph models. Further, are the expansion functions small enough to enable practical fpt algorithms? Preliminary empirical evaluations (using a related specialized coloring number) for random intersection graphs with $\alpha > 1$ indicate the answer is affirmative (details in the full version [11]). Furthermore, can the random intersection graph model be sensibly modified such that the clustering is tunable while being structurally sparse?

Acknowledgments. The authors thank Kevin Jasik of RWTH Aachen University for running the experiments in the full version of this paper. Portions of this research are a product of the ICERM research cluster "Towards Efficient Algorithms Exploiting Graph Structure", co-organized by B. Sullivan, E. Demaine, and D. Marx in April 2014.

N. Lemons funded by the Department of Energy at Los Alamos National Laboratory under contract DE-AC52-06NA25396 through the LDRD Program. F. Sánchez Villaamil funded by DFG-Project RO 927/13-1 "Pragmatic Parameterized Algorithms". B. D. Sullivan supported in part by DARPA GRAPHS/SPAWAR Grant N66001-14-1-4063, the Gordon & Betty Moore Foundation, and the National Consortium for Data Science. Any opinions, findings, and conclusions or recommendations expressed in this publication are those of the author(s) and do not necessarily reflect the views of DARPA, SSC Pacific, DOE, the Moore Foundation, or the NCDS.

References

1. Ball, F.G., Sirl, D.J., Trapman, P.: Epidemics on random intersection graphs. Ann. Appl. Probab. **24**(3), 1081–1128 (2014)
2. Blackburn, S.R., Stinson, D.R., Upadhyay, J.: On the complexity of the herding attack and some related attacks on hash functions. Des. Codes Crypt. **64**(1–2), 171–193 (2012)
3. Bloznelis, M.: Degree and clustering coefficient in sparse random intersection graphs. Ann. Appl. Probab. **23**, 1254–1289 (2013)
4. Bloznelis, M., Jaworski, J., Kurauskas, V.: Assortativity and clustering of sparse random intersection graphs. Electron. J. Probab. **18**, 1–24 (2013)
5. Bridson, M., Häfliger, A.: Metric Spaces of Non-Positive Curvature. Grundlehren Der Mathematischen Wissenschaften. Springer, Heidelberg (2009)
6. Chen, W., Fang, W., Hu, G., Mahoney, M.W.: On the hyperbolicity of small-world and tree-like random graphs. In: Chao, K.-M., Hsu, T., Lee, D.-T. (eds.) ISAAC 2012. LNCS, vol. 7676, pp. 278–288. Springer, Heidelberg (2012)
7. Chepoi, V., Dragan, F.F., Estellon, B., Habib, M., Vaxès, Y.: Diameters, centers, and approximating trees of δ-hyperbolic geodesic spaces and graphs. In: Symposium on Computational Geometry, pp. 59–68 (2008)
8. Deijfen, M., Kets, W.: Random intersection graphs with tunable degree distribution and clustering. Probab. Eng. Informational Sci. **23**, 661–674 (2009)
9. Demaine, E.D., Reidl, F., Rossmanith, P., Sánchez Villaamil, F., Sikdar, S., Sullivan, B.D.: Structural sparsity of complex networks: Bounded expansion in random models and real-world graphs. CoRR, abs/1406.2587 (2014)
10. Dvořák, Z., Král', D., Thomas, R.: Testing first-order properties for subclasses of sparse graphs. J. ACM **60**(5), 36 (2013)
11. Farrell, M., Goodrich, T., Lemons, N., Reidl, F., Sánchez Villaamil, F., Sullivan, B.D.: Hyperbolicity, degeneracy, and expansion of random intersection graphs. CoRR, abs/1409.8196 (2014)
12. Godehardt, E., Jarowski, J., Rybarczyk, K.: Clustering coefficients of random intersection graphs. In: Gaul, W.A., Geyer-Schulz, A., Schmidt-Thieme, L., Kunze, J. (eds.) Challenges at the Interface Of Data Analysis Computer Science And Optimization, pp. 243–253. Springer, Heidelberg (2012)
13. Gromov, M.: Hyperbolic groups. In: Gersten, S.M. (ed.) Essays in Group Theory. Mathematical Sciences Research Institute Publications, vol. 8, pp. 75–263. Springer, New York (1987)

14. Jaworski, J., Karoński, M., Stark, D.: The degree of a typical vertex in generalized random intersection graph models. Discrete Math. **306**, 2152–2165 (2006)
15. Jonckheere, E., Lohsoonthorn, P., Bonahon, F.: Scaled Gromov hyperbolic graphs. J. Graph Theory **57**(2), 157–180 (2008)
16. Karoński, M., Scheinerman, E.K., Singer-Cohen, K.B.: On random intersection graphs: the subgraph problem. Combin. Probab. Comput. **8**, 131–159 (1999)
17. Kleinberg, R.: Geographic routing using hyperbolic space. In: Proceedings of the 26th INFOCOM, pp. 1902–1909 (2007)
18. Narayan, O., Saniee, I.: Large-scale curvature of networks. Phys. Rev. E **84**, 066108 (2011)
19. Nešetřil, J., Ossona de Mendez, P.: Grad and classes with bounded expansion I. and II. Eur. J. Comb. **29**(3), 760–791 (2008)
20. Nešetřil, J., Ossona de Mendez, P.: First order properties on nowhere dense structures. J. Symbolic Logic **75**(3), 868–887 (2010)
21. Nešetřil, J., Ossona de Mendez, P.: On nowhere dense graphs. Eur. J. Comb. **32**(4), 600–617 (2011)
22. Nešetřil, J., Ossona de Mendez, P.: Sparsity: Graphs, Structures, and Algorithms. Algorithms and Combinatorics, vol. 28. Springer, Heidelberg (2012)
23. Nešetřil, J., Ossona de Mendez, P., Wood, D.R.: Characterisations and examples of graph classes with bounded expansion. Eur. J. Comb. **33**(3), 350–373 (2012)
24. Newman, M.E.J., Strogatz, S.H., Watts, D.J.: Random graphs with arbitrary degree distributions and their applications. Phys. Rev. E **64**(2), 026118 (2001)
25. Rybarczyk, K.: Diameter, connectivity, and phase transition of the uniform random intersection graph. Discrete Math. **311**, 1998–2019 (2011)
26. Rybarczyk, K.: The coupling method for inhomogeneous random intersection graphs. Preprint. arXiv:1301.0466 (2013)
27. Singer-Cohen, K.: Random intersection graphs. Ph.D. thesis, Department of Mathematical Sciences, The Johns Hopkins University (1995)
28. Watts, D.J., Strogatz, S.H.: Collective dynamics of 'small-world' networks. Nature **393**, 440–442 (1998)
29. Zhao, J.: Minimum node degree and k-connectivity in wireless networks with unreliable links. In: 2014 IEEE International Symposium on Information Theory (ISIT), pp. 246–250. IEEE (2014)
30. Zhao, J., Yagan, O., Gligor, V.: On k-connectivity and minimum vertex degree in random s-intersection graphs. In: ANALCO (2015)
31. Zhao, J., Yagan, O., Gligor, V.: Connectivity in secure wireless sensor networks under transmission constraints. In: 2014 52nd Annual Allerton Conference on Communication, Control, and Computing, pp. 1294–1301. IEEE (2014)
32. Zhao, J., Yagan, O., Gligor, V.: On the strengths of connectivity and robustness in general random intersection graphs. In: 2014 IEEE 53rd Annual Conference on Decision and Control (CDC), pp. 3661–3668. IEEE (2014)

Degree-Degree Distribution in a Power Law Random Intersection Graph with Clustering

Mindaugas Bloznelis[(✉)]

Vilnius University, Naugarduko 24, 03225 Vilnius, Lithuania
`mindaugas.bloznelis@mif.vu.lt`

Abstract. The bivariate distribution of degrees of adjacent vertices (degree-degree distribution) is an important network characteristic defining the statistical dependencies between degrees of adjacent vertices. We show the asymptotic degree-degree distribution of a sparse inhomogeneous random intersection graph and discuss its relation to the clustering and power law properties of the graph.

Keywords: Degree-degree distribution · Power law · Clustering coefficient · Random intersection graph · Affiliation network · Assortativity coefficient

1 Introduction

Correlations between degrees of adjacent vertices influence many network properties including the component structure, epidemic spreading, random walk performance, network robustness, etc., see [2,8,11,14,15] and references therein. The correlations are defined by the degree-degree distribution, i.e., the bivariate distribution of degrees of endpoints of a randomly chosen edge. In this paper we present an analytic study of the degree-degree distribution in a mathematically tractable random graph model of an affiliation network possessing tunable power law degree distribution and clustering coefficient. Our study is motivated by the interest in tracing the relation between the degree-degree distribution and clustering properties in a power law network.

Affiliation Network and Random Intersection Graph. An affiliation network defines adjacency relations between actors by using an auxiliary set of attributes. Let V denote the set of actors (nodes of the network) and W denote the auxiliary set of attributes. Every actor $v \in V$ is prescribed a collection $S_v \subset W$ of attributes and two actors $u, v \in V$ are declared adjacent in the network if they share some common attributes. For example one may interpret elements of W as weights and declare two actors adjacent whenever the total weight of shared attributes is above some threshold value. Here we consider the simplest case, where $u, v \in V$ are called adjacent whenever they share at least one common attribute, i.e., $S_u \cap S_v \neq \emptyset$. Two popular examples of real affiliation networks are the film actor network, where two actors are declared adjacent if

© Springer International Publishing Switzerland 2015
D.F. Gleich et al. (Eds.): WAW 2015, LNCS 9479, pp. 42–53, 2015.
DOI: 10.1007/978-3-319-26784-5_4

they have played in the same movie, and the collaboration network, where two scientists are declared adjacent if they have coauthored a publication.

A plausible model of a large affiliation network is obtained by prescribing the collections of attributes to actors at random [12,16]. Furthermore, in order to model the heterogeneity of human activity, every actor $v_j \in V$ is prescribed a random weight y_j reflecting its activity. Similarly, a random weight x_i is prescribed to each attribute $w_i \in W$ to model its attractiveness. Now an attribute $w_i \in W$ is included in the collection S_{v_j} at random and with probability proportional to the attractiveness x_i and activity y_j. In this way we obtain a random graph on the vertex set V sometimes called the inhomogeneous random intersection graph, see [5] and references therein.

We argue that the inhomogeneous random intersection graph can be considered as a realistic model of a power law affiliation network. Indeed, empirical evidence reported in [13] suggests that (at least in some social networks) the 'heavy-tailed degree distribution is causally determined by similarly skewed distribution of human activity'.

Rigorous Model. Let $X_1, \ldots, X_m, Y_1, \ldots, Y_n$ be independent non-negative random variables such that each X_i has the probability distribution P_1 and each Y_j has the probability distribution P_2. Given realized values $X = \{X_i\}_{i=1}^m$ and $Y = \{Y_j\}_{j=1}^n$ we define the random bipartite graph $H_{X,Y}$ with the bipartition $V \cup W$, where $V = \{v_1, \ldots, v_n\}$ and $W = \{w_1, \ldots, w_m\}$. Every pair $\{w_i, v_j\}$ is linked in $H_{X,Y}$ with probability

$$p_{ij} = \min\{1, \lambda_{ij}\}, \qquad \text{where} \quad \lambda_{ij} = \frac{X_i Y_j}{\sqrt{nm}},$$

independently of the other pairs $\{w, v\} \in W \times V$. The inhomogeneous random intersection graph $G = G(P_1, P_2, n, m)$ defines the adjacency relation on the vertex set V: vertices $u, v \in V$ are declared adjacent (denoted $u \sim v$) whenever u and v have a common neighbor in $H_{X,Y}$. We call this neighbor a witness of the adjacency relation $u \sim v$.

The random graph G has several features that make it a convenient theoretical model of a real complex network. Firstly, the statistical dependence of neighboring adjacency relations in G mimics that of real affiliation networks. In particular, G admits a tunable clustering coefficient: For $m/n \to \beta \in (0, +\infty)$ as $m, n \to +\infty$, we have, see [7],

$$\mathbf{P}(v_1 \sim v_2 | v_1 \sim v_3, v_2 \sim v_3) = \frac{\kappa}{\kappa + \sqrt{\beta}} + o(1). \tag{1}$$

Here $\kappa := b_1 b_2^{-1} a_3 a_2^{-2}$ and $a_i = \mathbf{E}X_1^i$, $b_j = \mathbf{E}Y_1^j$.

Another important feature of the model is its ability to produce a rich class of (asymptotic) degree distributions including power law distributions. The following result of [4] will be used below. $d(v)$ denotes the degree of a vertex $v \in V$ in G.

Theorem 1. *Let $m, n \to \infty$.*

(i) Assume that $m/n \to \beta$ for some $\beta \in (0, +\infty)$. Suppose that $\mathbf{E}X_1^2 < \infty$ and $\mathbf{E}Y_1 < \infty$. Then $d(v_1)$ converges in distribution to the random variable

$$d_* = \sum_{j=1}^{\Lambda_1} \tau_j, \qquad (2)$$

where $\tau_1, \tau_2, \ldots$ are independent and identically distributed random variables independent of the random variable Λ_1. They are distributed as follows. For $r = 0, 1, 2, \ldots,$ we have

$$\mathbf{P}(\tau_1 = r) = \frac{r+1}{\mathbf{E}\Lambda_0}\mathbf{P}(\Lambda_0 = r+1) \quad \text{and} \quad \mathbf{P}(\Lambda_i = r) = \mathbf{E}\,e^{-\lambda_i}\frac{\lambda_i^r}{r!}, \quad i = 0, 1.$$
$$(3)$$

Here $\lambda_0 = X_1 b_1 \beta^{-1/2}$ and $\lambda_1 = Y_1 a_1 \beta^{1/2}$.

(ii) Assume that $m/n \to +\infty$. Suppose that $\mathbf{E}X_1^2 < \infty$ and $\mathbf{E}Y_1 < \infty$. Then $d(v_1)$ converges in distribution to a random variable Λ_3 having the probability distribution

$$\mathbf{P}(\Lambda_3 = r) = \mathbf{E}\,e^{-\lambda_3}\frac{\lambda_3^r}{r!}, \qquad r = 0, 1, \ldots. \qquad (4)$$

Here $\lambda_3 = Y_1 a_2 b_1$.

(iii) Assume that $m/n \to 0$. Suppose that $\mathbf{E}X_1 < \infty$. Then $\mathbf{P}(d(v_1) = 0) = 1 - o(1)$.

Using the fact that a Poisson random variable is highly concentrated around its mean one can show that for a power law distribution $\mathbf{P}(\lambda_i > r) \sim c_i\,r^{-\varkappa_i}$, with some $c_i, \varkappa_i > 0$, we have $\mathbf{P}(\Lambda_i > r) \sim c_i r^{-\varkappa_i}$, for $i = 0, 1, 3$. Here and below for real sequences $\{t_r\}_{r \geq 1}$ and $\{s_r\}_{r \geq 1}$ we denote $t_r \sim s_r$ whenever $t_r/s_r \to 1$ as $r \to +\infty$. Furthermore, the tail $\mathbf{P}(d^* > r)$ of a randomly stopped sum d_* is as heavy as the heavier one of Y_1 and τ_1, see, e.g., [1]. Hence, choosing a power law weights X and Y we obtain a power law asymptotic degree distributions, namely, the distributions of d^* and Λ_3.

In what follows we will focus on the local probabilities. Given $c > 0$ and $\varkappa > 1$, we say that a non-negative random variable Z has the power law property $\mathcal{P}_{c,\varkappa}$ (denoted $Z \in \mathcal{P}_{c,\varkappa}$) if either Z is integer valued and satisfies $\mathbf{P}(Z = r) \sim cr^{-\varkappa}$, or Z is absolute continuous with density f_Z satisfying $f_Z(t) = (c + o(1))t^{-\varkappa}$ as $t \to +\infty$.

Several examples are considered in Remark 1.

Remark 1. *Let $c, \varkappa > 0$. Let $m, n \to +\infty$.*

(i) Let $a > 0$ and $\varkappa > 3$. Assume that $\mathbf{E}e^{aY_1} < \infty$ and $X_1 \in \mathcal{P}_{c,\varkappa}$. Then $\mathbf{P}(d_* = r) \sim cb_1^{\varkappa-1}\beta^{(3-\varkappa)/2}r^{1-\varkappa}$.

(ii) Let $\varkappa > 2$. Assume that $Y_1 \in \mathcal{P}_{c,\varkappa}$ and $\mathbf{P}(X_1 = x) = 1$. Then $\mathbf{P}(d_* = r) \sim c(x^2 b_1)^{\varkappa-1}r^{-\varkappa}$.

The proof of (i) makes use of power law properties of the local probabilities of randomly stopped sums, like d_*, in the case where the summands are heavy tailed and the number of summands has a light tail (see, e.g., Theorem 4.30 of [9]). Unfortunatelly we are not aware of rigorous results establishing power law properties of the local probabilities of randomly stopped sums in the case where the *number of summands is heavy tailed.*

Degree-Degree Distribution. We are interested in the bivariate distribution of degrees of adjacent vertices. Denote $d_1 = d(v_1)$, $d_2 = d(v_2)$ and let

$$p(k_1, k_2) = \mathbf{P}(d_1 = k_1 + 1, d_2 = k_2 + 1 \mid v_1 \sim v_2), \qquad k_1, k_2 = 0, 1, \ldots, \quad (5)$$

denote the probabilities defining the conditional bivariate distribution of the *ordered* pair (d_1, d_2), given the event that vertices v_1 and v_2 are adjacent. Let (u^*, v^*) be an ordered pair of distinct vertices chosen uniformly at random from V. We note that

$$p(k_1, k_2) = p(k_2, k_1) = \mathbf{P}\big(d(u^*) = k_1 + 1, d(v^*) = k_2 + 1 \mid u^* \sim v^*\big),$$

since the probability distribution of graph G is invariant under permutations of its vertices.

In Theorem 2 we collect several examples of degree-degree distributions admitting explicit asymptotic formulas.

Theorem 2. *Let $c, x > 0$. Let $m, n \to +\infty$.*

(i) Assume that $\mathbf{E}X_1^2 < \infty$. Let $\varkappa > 2$ and assume that $Y_1 \in \mathcal{P}_{c,\varkappa}$. Suppose that $m/n \to +\infty$. Then for $k_1, k_2 \to +\infty$ we have

$$p(k_1, k_2) = (1 + o(1))c^2 a_2^{2\varkappa - 4} b_1^{2\varkappa - 6} (k_1 k_2)^{1 - \varkappa}. \quad (6)$$

(ii) Let $a, \beta > 0$ and $\varkappa > 3$. Assume that $\mathbf{E}e^{aY_1} < \infty$ and $X_1 \in \mathcal{P}_{c,\varkappa}$. Suppose that $m/n \to \beta$. Let $k_1, k_2 \to +\infty$ so that $k_1 \leq k_2$. Suppose that either $k_2 - k_1 \to +\infty$ or $k_2 - k_1 = k$, for an arbitrary, but fixed integer $k \geq 0$. Then

$$p(k_1, k_2) = (1 + o(1)) \frac{\beta}{b_1^4 a_2} c_1^* f(k_1, k_2), \quad (7)$$

where

$$f(k_1, k_2) = c_2^* k_1^{2 - \varkappa} (k_2 - k_1)^{1 - \varkappa}, \qquad \text{for} \quad k_2 - k_1 \to +\infty,$$
$$f(k_1, k_2) = c_{3,k}^* k_1^{2 - \varkappa}, \qquad\qquad \text{for} \quad k_2 - k_1 = k.$$

Here $c_1^* = c(b_1 \beta^{-1/2})^{\varkappa - 1}$, $c_2^* = cb_1^{\varkappa - 2} b_2 \beta^{(3 - \varkappa)/2}$, and $c_{3,k}^* = \sum_{i \geq 0} q_i q_{k+i}$. Furthermore, we have $q_r \sim c_2^* r^{1 - \varkappa}$. We note that q_j are defined in (9).

(iii) Let $\varkappa > 2$ and $\beta > 0$. Assume that $Y_1 \in \mathcal{P}_{c,\varkappa}$ and $\mathbf{P}(X_1 = x) = 1$. Suppose that $m/n \to \beta$. For $k_1, k_2 \to +\infty$ we have

$$p(k_1, k_2) = (1 + o(1))c^2 x^{4\varkappa - 8} b_1^{2\varkappa - 6} (k_1 k_2)^{1 - \varkappa}. \quad (8)$$

The assumption $m/n \to +\infty$ of example (i) of Theorem 2 implies that the adjacency relations in G are asymptotically independent as $m, n \to +\infty$. In this case the clustering coefficient vanishes and the asymptoptic degree-degree distribution is the product of size biased asymptotic degree distributions, see Theorem 3 and Remark 2 below.

The assumption $m/n \to \beta < \infty$ of examples (ii) and (iii) of Theorem 2 implies that the adjacency relations remain statistically dependent as $m, n \to +\infty$. In this case G admits a non-vanishing clustering and assortativity coefficients (degrees of adjacent vertices are positively correlated), [6]. A combination of heavy tailed weights X and light weights Y of example (ii) produces quite a complex pattern of the collection of bivariate probabilities $p(k_1, k_2)$. It seems a bit surprising to us that a combination of light weights X and heavy tailed Y of example (iii) of Theorem 2 create bivariate probabilities with asymptotically independent marginals for $k_1, k_2 \to +\infty$.

Theorem 2 is a corollary of a general result stated in Theorem 3 below. Before formulation of Theorem 3 we introduce some notation.

We remark that d_* defined by (2) depends on Y_1. By conditioning on the event $\{Y_1 = y\}$ we obtain another random variable, denoted d_y^*, which has the compound Poisson distribution

$$\mathbf{P}(d_y^* = r) = \mathbf{P}(d_* = r | Y_1 = y) = \mathbf{P}\Big(\sum_{j=1}^{N} \tau_j = r\Big), \qquad r = 0, 1, \dots.$$

Here $N = N_y$ denotes a Poisson random variable which is independent of the iid sequence $\{\tau_j\}_{j \geq 1}$ and has mean $\mathbf{E}N_y = ya_1\beta^{1/2}$, $y \geq 0$. Given integers $k_1, k_2, r \geq 0$, denote

$$q_r = \mathbf{E}\big(Y_1 \mathbf{P}(d_{Y_1}^* = r | Y_1)\big) = \mathbf{E}\big(Y_1 \mathbf{P}(d_* = r | Y_1)\big), \tag{9}$$

$$p_\beta(k_1, k_2) = \frac{\beta}{b_1^4 a_2} \sum_{r=0}^{k_1 \wedge k_2} (r+1)(r+2)\mathbf{P}(\Lambda_0 = r+2)q_{k_1-r}q_{k_2-r},$$

$$\tilde{p}(r) = r\mathbf{P}(\Lambda_3 = r)\big(\mathbf{E}\Lambda_3\big)^{-1}, \qquad p_\infty(k_1, k_2) = \tilde{p}(k_1+1)\tilde{p}(k_2+1).$$

Our main result is the following theorem.

Theorem 3. *Let $m, n \to \infty$. Suppose that $\mathbf{E}X_1^2 < \infty$ and $\mathbf{E}Y_1 < \infty$.*

(i) Assume that $m/n \to \beta$ for some $\beta \in (0, +\infty)$. Then for every $k_1, k_2 \geq 0$ we have

$$p(k_1, k_2) = p_\beta(k_1, k_2) + o(1), \tag{10}$$

(ii) Assume that $m/n \to +\infty$. Then for every $k_1, k_2 \geq 0$ we have

$$p(k_1, k_2) = p_\infty(k_1, k_2) + o(1). \tag{11}$$

We note that the moment conditions $\mathbf{E}X_1^2 < \infty$ and $\mathbf{E}Y_1 < \infty$ of Theorem 3 are the minimal ones as $a_2 = \mathbf{E}X_1^2$ and $b_1 = \mathbf{E}Y_1$ enter the expressions defining the quantities on the right side of (10) and (11).

Remark 2. *In the case where $m/n \to +\infty$, the size biased probability distribution $\{\tilde{p}(r)\}_{r \geq 1}$ is the limiting distribution of $d(v_1)$ conditioned on the event $v_1 \sim v_2$, i.e.,*

$$\mathbf{P}(d(v_1) = r | v_1 \sim v_2) = \tilde{p}(r) + o(1), \qquad r = 1, 2, \dots. \tag{12}$$

Our final remark is about the case where $m/n \to 0$. By Theorem 1, in this case the edges of a sparse inhomogeneous random intersection graph span a subgraph on $o(n)$ randomly selected vertices leaving the remaining $(1 - o(1))n$ vertices isolated. Consequently, the subgraph is relatively dense and we do not expect stochastically bounded degrees of endpoints of adjacent vertices.

Related Work. The influence of degree-degree correlations on the network properties have been studied by many authors, see, e.g., [8,11,14,15] and references therein. The asymptotic degree-degree distribution in a preferential attachment random graph with tunable power law degree distribution was shown in [10]. Our model and approach are much different compared to [10]. To our best knowledge the present paper is the first attempt to trace the relation between the degree-degree distribution and the *clustering property* in a power law network. Connections between Newman's assortativity coefficient and the clustering coefficient in a related random graph model has been discussed in [6].

2 Proofs

The section is organized as follows. Before the proofs we present auxiliary Lemma 1. Then we prove Remark 1, Theorem 2 and sketch the Proof of Theorem 3. The complete Proof of Theorem 3 is given in [3]. At the end of the section we prove Remark 2.

Lemma 1. *Let $c, \varkappa, h > 0$. Let Z, Λ_Z be non-negative random variables such that $\mathbf{P}(\Lambda_Z = r) = \mathbf{E}(e^{-Z} Z^r / r!)$, $r = 0, 1, \ldots$.*
(i) The relation $\mathbf{P}(Z > t) = (c + o(1))t^{-\varkappa}$ as $t \to +\infty$ implies

$$\mathbf{P}(\Lambda_Z > t) = (c + o(1))t^{-\varkappa} \qquad as \quad t \to +\infty. \tag{13}$$

(ii) Assume that $Z \in \mathcal{P}_{c,\varkappa}$. Then $\mathbf{P}(\Lambda_Z = r) \sim cr^{-\varkappa}$.
(iii) If hZ is integer valued and satisfies $\mathbf{P}(hZ = r) \sim c(h/r)^{\varkappa}$ then $\mathbf{P}(\Lambda_Z = r) \sim chr^{-\varkappa}$.

The Proof of Lemma 1 is technical. It is given in [3].

Proof of Remark 1. Let us prove (i). Lemma 1 implies

$$\mathbf{P}(\Lambda_0 = r) \sim c_1^* r^{-\varkappa}, \qquad where \qquad c_1^* = c(b_1 \beta^{-1/2})^{\varkappa - 1}. \tag{14}$$

Consequently,

$$\mathbf{P}(\tau_1 = r) \sim c a_1^{-1} (b_1 \beta^{-1/2})^{\varkappa - 2} r^{1 - \varkappa}. \tag{15}$$

From the latter relation we conclude that the sequence of probabilities $\{\mathbf{P}(\tau_1 = r)\}_{r \geq 0}$ is longtailed and subexponential, that is, it satisfies conditions of Theorem 4.30 of [9]. This theorem implies $\mathbf{P}(\sum_{1 \leq i \leq \Lambda_1} \tau_i = r) \sim \mathbf{P}(\tau_1 = r)\mathbf{E}\Lambda_1$, thus completing the proof.

Let us prove (ii). We observe that τ_1 has Poisson distribution with mean $\lambda_0 = \varkappa b_1 \beta^{-1/2}$. Hence, given Λ_1, the random variable d_* has Poisson distribution with mean $\lambda_0 \Lambda_1$. Now statement (iii) of Lemma 1 implies that $\mathbf{P}(\Lambda_1 = r) \sim c_* r^{-\varkappa}$, where $c_* = c(a_1 \beta^{1/2})^{\varkappa - 1}$. Next, we apply statement (iii) of Lemma 1 once again and obtain $\mathbf{P}(d_* = r) \sim c_* \lambda_0^{\varkappa - 1} r^{-\varkappa}$.

Proof of Theorem 2. Statement (i) follows from the relation $\mathbf{P}(\Lambda_3 = r) \sim c(a_2 b_1)^{\varkappa-1} r^{-\varkappa}$, see Lemma 1.

In the proof of (ii) and (iii) we assume that $k_1 \leq k_2$ and use the notation

$$S_A = \sum_{r \in A} (r+1)(r+2)\mathbf{P}(\Lambda_0 = r+2)q_{k_1-r}q_{k_2-r}, \qquad A \in [0, k_1].$$

Let us prove (ii). We observe that $\mathbf{E}(e^{aY_1}) < \infty$ implies that $\mathbf{E}Y_1 e^{a'\Lambda_1} < \infty$ for some $a' > 0$. Using this observation and the fact that the sequence of probabilities $\{\mathbf{P}(\tau_1 = r)\}_{r \geq 0}$ is longtailed and subexponential (see (15) and [9]) we show that

$$\mathbf{E}\big(Y_1\mathbf{P}(d^*_{Y_1} = r|Y_1)\big) \sim \big(\mathbf{E}(Y_1\Lambda_1)\big)\mathbf{P}(\tau_1 = r). \tag{16}$$

The proof of (16) is much the same as that of Theorem 4.30 in [9]. Now, we invoke in (16) the identity $\mathbf{E}(Y_1\Lambda_1) = \mathbf{E}(Y_1\lambda_1) = a_1 b_2 \beta^{1/2}$ and (15), and obtain

$$\mathbf{E}\big(Y_1\mathbf{P}(d^*_{Y_1} = r|Y_1)\big) \sim c^*_2 r^{1-\varkappa}, \qquad \text{where} \qquad c^*_2 = cb_1^{\varkappa-2} b_2 \beta^{(3-\varkappa)/2}.$$

Hence we have $q_r \sim c^*_2 r^{1-\varkappa}$ and $\mathbf{P}(\Lambda_0 = r) \sim c^*_1 r^{-\varkappa}$, see (14).

Now we are ready to prove (7). Let $\varepsilon = \ln(k_1 \wedge (k_2 - k_1))$ for $k_2 - k_1 \to +\infty$, and $\varepsilon = \ln k_1$ otherwise. Split $S_{[0,k_1]} = S_{A_1} + S_{A_2} + S_{A_3}$, where

$$A_1 = [0, k_1/2], \qquad A_2 = (k_1/2, k_1 - \varepsilon], \qquad A_3 = (k_1 - \varepsilon, k_1].$$

In the remaining part of the proof we shall show that S_{A_1}, S_{A_2} are negligibly small compared to S_{A_3} and determine the first order asymptotics of S_{A_3} as $k_1, k_2 \to +\infty$. We have for some $\bar{c} > 0$ (independent of k_1, k_2)

$$S_{A_1} \leq \bar{c} \sum_{i \in A_1} \frac{1}{(k_1 - i)^{\varkappa-1}} \frac{1}{(k_2 - i)^{\varkappa-1}} \frac{1}{(1+i)^{\varkappa-2}} = O\Big(k_1^{4-2\varkappa} k_2^{1-\varkappa}(1+\Delta)\Big).$$

Here $\Delta = \ln n$ for $\varkappa = 3$ and $\Delta = 0$ otherwise. Furthermore, for $k_2 - k_1$ bounded we have

$$S_{A_2} \leq \frac{\bar{c}}{k_1^{\varkappa-2}} \sum_{i \in A_2} \frac{1}{(k_1 - i)^{2\varkappa-2}} = O\Big(\varepsilon^{3-2\varkappa} k_1^{2-\varkappa}\Big) = o(k_1^{2-\varkappa}).$$

For $k_2 - k_1 \to +\infty$ we have

$$S_{A_2} \leq \frac{\bar{c}}{k_1^{\varkappa-2}} \sum_{i \in A_2} \frac{1}{(k_1 - i)^{\varkappa-1}} \frac{1}{(k_2 - i)^{\varkappa-1}} \leq \frac{\bar{c}}{k_1^{\varkappa-2}(k_2 - k_1)^{\varkappa-1}} \sum_{i \in A_2} \frac{1}{(k_1 - i)^{\varkappa-1}}.$$

Since $\sum_{i \in A_2} \frac{1}{(k_1-i)^{\varkappa-1}} = O(\varepsilon^{2-\varkappa}) = o(1)$ we obtain $S_{A_2} = o\big(k_1^{2-\varkappa}(k_2 - k_1)^{1-\varkappa}\big)$.

Finally, using the approximation $(i+1)(i+2)\mathbf{P}(\Lambda_0 = i+2) \sim c^*_1 k_1^{2-\varkappa}$ uniformly in $i \in A_3$ we obtain for $k_2 - k_1 \to +\infty$

$$S_{A_3} = \frac{c^*_1(1+o(1))}{k_1^{\varkappa-2}} \sum_{i \in A_3} q_{k_1-i} q_{k_2-i} \sim \frac{c^*_1}{k_1^{\varkappa-2}} \frac{c^*_2}{(k_2 - k_1)^{\varkappa-1}} \sum_{i \in A_3} q_{k_1-i}$$

$$\sim \frac{c^*_1 c^*_2}{k_1^{\varkappa-2}(k_2 - k_1)^{\varkappa-1}}.$$

Similarly, in the case where $k_2 - k_1 = k$ for some fixed k we have

$$S_{A_3} = \frac{c_1^*(1 + o(1))}{k_1^{\varkappa - 2}} \sum_{i \in A_3} q_{k_1 - i} q_{k_2 - i} \sim \frac{c_1^* c_{3,k}^*}{k_1^{\varkappa - 2}}, \quad \text{where} \quad c_{3,k}^* = \sum_{i \geq 0} q_i q_{k+i}.$$

Let us prove (iii). We observe that Λ_0 has the Poisson distribution with (non-random) mean λ_0. Using the identity $(r+1)(r+2)\mathbf{P}(\Lambda_0 = r+2) = \lambda_0^2 \mathbf{P}(\Lambda_0 = r)$ we write

$$S_{[0,k_1]} = \lambda_0^2 \sum_{r=0}^{k_1} \mathbf{P}(\Lambda_0 = r) q_{k_1 - r} q_{k_2 - r} = \lambda_0^2 \mathbf{E}(q_{k_1 - \Lambda_0} q_{k_2 - \Lambda_0}) = \lambda_0^2 (J_1 + J_2),$$

$$J_1 = \mathbf{E}(q_{k_1 - \Lambda_0} q_{k_2 - \Lambda_0}) \mathbb{I}_{\{\Lambda_0 < \sqrt{k_1}\}}, \qquad J_2 = \mathbf{E}(q_{k_1 - \Lambda_0} q_{k_2 - \Lambda_0}) \mathbb{I}_{\{\sqrt{k_1} \leq \Lambda_0 \leq k_1\}}.$$

Next, combining the fast decay of Poisson tail probability $\mathbf{P}(\Lambda_0 > t)$ as $t \to +\infty$ with the relation, which is shown below,

$$q_r \sim c_0 r^{1-\varkappa}, \qquad c_0 = c(x^2 b_1)^{\varkappa - 2}, \tag{17}$$

we estimate $J_1 = (1 + o(1)) c_0^2 (k_1 k_2)^{1-\varkappa}$ and $J_2 = o((k_1 k_2)^{1-\varkappa}$. We obtain that $S_{[0,1]} = (1 + o(1)) \lambda_0^2 J_1$. Now the identity $p_\beta(k_1, k_2) = \beta b_1^{-4} x^{-2} S_{[0,k_1]}$ completes the proof of (8).

Now, we prove (17). Since τ_1 has Poisson distribution with mean $\lambda_0 = x b_1 \beta^{-1/2}$, we obtain

$$\mathbf{P}(d_y^* = k) = \mathbf{E}\left(e^{-\lambda_0 \Lambda_1} \frac{(\lambda_0 \Lambda_1)^k}{k!} \Big| Y_1 = y\right) = \sum_{i \geq 0} e^{-\lambda_0 i} \frac{(\lambda_0 i)^k}{k!} e^{-By} \frac{(By)^i}{i!}.$$

Here we denote $B = x \beta^{1/2}$. After we write the product $y \mathbf{P}(d_y^* = k)$ in the form

$$\sum_{i \geq 0} e^{-\lambda_0 i} \frac{(\lambda_0 i)^k}{k!} e^{-By} \frac{(By)^{i+1}}{(i+1)!} \frac{i+1}{B}$$

$$= \mathbf{E}\left(e^{-\lambda_0 (\Lambda_1 - 1)} \frac{(\lambda_0 (\Lambda_1 - 1))^k}{k!} \frac{\Lambda_1}{B} \mathbb{I}_{\{\Lambda_1 \geq 1\}} \Big| Y_1 = y\right),$$

we obtain the following expression for the expectation $q_k = \mathbf{E}(Y_1 \mathbf{P}(d_{Y_1}^* = k|Y_1))$:

$$q_k = \frac{\mathbf{P}(\Lambda_1 \geq 1)}{B} (I_{1,k} + I_{2,k}), \qquad I_{1,k} = \mathbf{E}\left(e^{-\lambda_0 Z} \frac{(\lambda_0 Z)^k}{k!}\right), \tag{18}$$

$$I_{2,k} = \mathbf{E}\left(Z e^{-\lambda_0 Z} \frac{(\lambda_0 Z)^k}{k!}\right).$$

Here Z denotes a random variable with the distribution

$$\mathbf{P}(Z = r) = \mathbf{P}(\Lambda_1 = r + 1) / \mathbf{P}(\Lambda_1 \geq 1), \qquad r = 0, 1, \dots.$$

We note that

$$\mathbf{P}(Z = r) \sim c' r^{-\varkappa}, \qquad \text{where} \qquad c' = cB^{\varkappa-1}/\mathbf{P}(\Lambda_1 \geq 1). \qquad (19)$$

Indeed, (19) follows from the relation $\mathbf{P}(\Lambda_1 = r) \sim cB^{\varkappa-1}r^{-\varkappa}$, which is a simply consequence of the property $\mathcal{P}_{c,\varkappa}$ of the distribution of Y_1, see Lemma 1. Next, we show that

$$I_{1,k} \sim c'\lambda_0^{\varkappa-1}k^{-\varkappa}, \qquad I_{2,k} \sim c'\lambda_0^{\varkappa-2}k^{1-\varkappa}. \qquad (20)$$

The first relation follows from (19), by Lemma 1. The second relation follows from the first one via the simple identity $I_{2,k} = (k+1)\lambda_0^{-1}I_{1,k+1}$. Finally invoking (20) in (18) we obtain (17).

Proof of Theorem 3. Before the proof we collect some notation.

Given two real sequences $\{a_n\}_{n\geq 1}$ and $\{b_n\}_{n\geq 1}$ we write $a_n \approx b_n$ to denote the fact that $(a_n - b_n)mn = o(1)$. By $\mathbb{I}_{\mathcal{A}}$ we denote the indicator function of an event $\mathcal{A}$. $\bar{\mathbb{I}}_{\mathcal{A}} = 1 - \mathbb{I}_{\mathcal{A}} = \mathbb{I}_{\bar{\mathcal{A}}}$ denotes the indicator of the complement event $\bar{\mathcal{A}}$.

The number of common neighbors of $v_i, v_j \in V$ is denoted d_{ij}. For a vertex $v \in V$ and attribute $w \in W$ we denote by $\{w \to v\}$ the event that v and w are linked in H. Introduce the events $\mathcal{A}_i = \{w_i \to v_1, w_i \to v_2\}$, $1 \leq i \leq m$. We write for short

$$\mathbb{I}_{ij} = \mathbb{I}_{\{w_i \to v_j\}}, \qquad \bar{\mathbb{I}}_{ij} = 1 - \mathbb{I}_{ij}, \qquad \mathbf{I}_{ij} = \mathbb{I}_{\{\lambda_{ij} \leq 1\}}, \qquad \bar{\mathbf{I}}_{ij} = 1 - \mathbf{I}_{ij}.$$

Let $\mathbf{L} = (L_0, L_1, L_2)$ denote the random vector with marginal random variables

$$L_0 = u_1, \quad L_1 = \sum_{2 \leq i \leq m} \mathbb{I}_{i1}u_i, \quad L_2 = \sum_{2 \leq i \leq m} \mathbb{I}_{i2}u_i, \quad u_i = \sum_{3 \leq j \leq n} \mathbb{I}_{ij}, \quad 1 \leq i \leq m.$$

Let $\Lambda_0, \Lambda_1, \Lambda_2, \Lambda_3, \Lambda_4$ denote random variables having mixed Poisson distributions

$$\mathbf{P}(\Lambda_i = r) = \mathbf{E}(e^{-\lambda_i}\lambda_i^r/r!), \qquad r = 0, 1, 2, \ldots, \quad i = 0, 1, 2, 3, 4. \qquad (21)$$

Here

$$\lambda_0 = X_1 b_1 \beta^{-1/2}, \quad \lambda_1 = Y_1 a_1 \beta^{1/2}, \quad \lambda_2 = Y_2 a_1 \beta^{1/2}, \quad \lambda_3 = Y_1 a_2 b_1, \quad \lambda_4 = Y_2 a_2 b_1.$$

We assume that conditionally, given Y_1, Y_2, X_1, the random variables Λ_i, $0 \leq i \leq 4$ are independent. Define random variables $d_{Y_1}^* = \sum_{i=1}^{\Lambda_1} \tau_i$ and $d_{Y_2}^* = \sum_{i=1}^{\Lambda_2} \tau_i'$. Here $\tau_i, \tau_i', i \geq 1$ are independent and identically distributed random variables, which are independent of Y_1, Y_2, X_1 and have distribution (3). Define the events

$$\mathcal{U}_{k_1,k_2} = \{d_1 = k_1 + 1, d_2 = k_2 + 1\}, \qquad \mathcal{U}_{r,r_1,r_2}^* = \{\mathbf{L} = (r, r_1, r_2)\},$$
$$\mathcal{U}_{r,r_1,r_2}^{**} = \{\Lambda_0 = r, d_{Y_1}^* = r_1, d_{Y_2}^* = r_2\}, \qquad \mathcal{U}_{r_1,r_2}^{***} = \{\Lambda_3 = r_1, \Lambda_4 = r_2\}.$$

Proof of (i). The intuition behind formula (10) is that the adjacency relation $v_1 \sim v_2$ as well as the common neighbors of v_1 and v_2 are witnessed, with a high probability, by a single attribute (all attributes having equal chances). Furthermore, conditionally on the event that this attribute is w_1, and given Y_1, Y_2, X_1, we have that the random variables d_{12}, $d_1 - 1 - d_{12} =: d_1'$, $d_2 - 1 - d_{12} =: d_2'$ are asymptotically independent. We note that d_1' and d_2' count individual (not shared) neighbors of v_1 and v_2. The asymptotic independence comes from the fact that (conditionally given Y_1, Y_2, X_1) these random variables are mainly related via average characteristics $m^{-1} \sum_{2 \leq i \leq m} X_i^j$ and $n^{-1} \sum_{3 \leq i \leq n} Y_i^j$, which are asymptotically constant, by the law of large numbers. Now, using Theorem 1 we identify limiting distributions of d_1', d_2'. The limiting distribution of d_{12} (conditioned on the event $\mathcal{A}_1$) is that of Λ_0.

We briefly outline the proof. Firstly, combining the identity $\{v_1 \sim v_2\} = \cup_{i=1}^m \mathcal{A}_i$ with approximations

$$\mathbf{P}(\cup_i \mathcal{A}_i) = \sum_i \mathbf{P}(\mathcal{A}_i) + o(n^{-1}) \quad \text{and} \quad \mathbf{P}(\mathcal{A}_i) \approx a_2 b_1^2 / (mn)$$

we show that $\mathbf{P}(v_1 \sim v_2) = a_2 b_1^2 n^{-1} + o(n^{-1})$ and

$$p(k_1, k_2) = \frac{\mathbf{P}(\mathcal{U}_{k_1, k_2} \cap \{v_1 \sim v_2\})}{\mathbf{P}(v_1 \sim v_2)} = \frac{\mathbf{P}(\mathcal{U}_{k_1, k_2} \cap (\cup_i \mathcal{A}_i))}{\mathbf{P}(\cup_i \mathcal{A}_i)}$$

$$= \frac{nm}{a_2 b_1^2} \mathbf{P}(\mathcal{U}_{k_1, k_2} \cap \mathcal{A}_1) + o(1). \tag{22}$$

Then using the total probability formula we split

$$\mathbf{P}(\mathcal{U}_{k_1, k_2} \cap \mathcal{A}_1) = \sum_{r=0}^{k_1 \wedge k_2} \mathbf{P}(\mathcal{U}_{k_1, k_2} \cap \{d_{12} = r\} \cap \mathcal{A}_1). \tag{23}$$

Secondly, we approximate random variables d_{12}, d_1', d_2' by L_0, L_1, L_2, and show that for every $r = 0, 1, \ldots, k_1 \wedge k_1$

$$\mathbf{P}(\mathcal{U}_{k_1, k_2} \cap \{d_{12} = r\} \cap \mathcal{A}_1) \approx \mathbf{P}(\mathcal{U}_{r, k_1 - r, k_2 - r}^* \cap \mathcal{A}_1). \tag{24}$$

Finally, using the Poisson approximation we approximate the sums of random indicators L_0, L_1, L_2 by independent random variables $\Lambda_0, d_{Y_1}^*, d_{Y_2}^*$. In this way we obtain the approximations, for $r, r_1, r_2 = 0, 1, 2 \ldots,$

$$\mathbf{P}(\mathcal{U}_{r, r_1, r_2}^* \cap \mathcal{A}_1) \approx \mathbf{E}\big(\lambda_{11} \lambda_{12} \mathbf{P}(\mathcal{U}_{r, r_1, r_2}^{**} | X_1, Y_1, Y_2)\big). \tag{25}$$

Now the simple identity

$$nm \mathbf{E}\big(\lambda_{11} \lambda_{12} \mathbf{P}(\mathcal{U}_{r, r_1, r_2}^{**} | X_1, Y_1, Y_2)\big) = \frac{\beta}{b_1^2}(r+1)(r+2)\mathbf{P}(\Lambda_0 = r + 2)q_{r_1} q_{r_2}$$

completes the proof of (10).

Proof of (ii). The proof (11) is similar to that of (10). It makes use the observation that the typical adjacency relation is witnessed by a single attribute. One difference is that now the size of the collection of attributes, prescribed to the typical vertex, tends to infinity as $n, m \to +\infty$. Since our intersection graph is sparse, this implies that the number of vertices linked (in H) to any given single attribute is most likely 0. This number is 1 with a small probability, but it is almost never larger than 1. As a consequence we obtain that $d_{12} = o_P(1)$. Another consequence is that the number of neighbors of a given vertex is distributed as Poisson mixture $\mathcal{P}(\lambda)$, where random variable λ accounts for the size of the collection of attributes prescribed to the vertex.

The first several steps of the proof are the same as that of (10). Namely, relations (22), (23), (24) hold true as their proof remains valid for $m/n \to +\infty$. Further steps of the proof are a bit different. We show that

$$\mathbf{P}(\mathcal{U}^*_{r,k_1-r,k_2-r} \cap \mathcal{A}_1) \approx 0, \qquad \text{for} \qquad r = 1, 2, \ldots, \tag{26}$$

and

$$\mathbf{P}(\mathcal{U}^*_{0,k_1,k_2} \cap \mathcal{A}_1) \approx \mathbf{P}(\{L_1 = k_1, L_2 = k_2\} \cap \mathcal{A}_1). \tag{27}$$

Finally, using the Poisson approximation we approximate the sums of indicators L_1, L_2 by Poisson mixtures Λ_3, Λ_4. In this way we obtain the approximation

$$\mathbf{P}(\{L_1 = k_1, L_2 = k_2\} \cap \mathcal{A}_1) \approx \mathbf{E}\big(\lambda_{11}\lambda_{12}\mathbf{P}(\mathcal{U}^{***}_{k_1,k_2}|Y_1, Y_2)\big). \tag{28}$$

Now the simple identity

$$a_2^{-1}b_1^{-2}nm\mathbf{E}\big(\lambda_{11}\lambda_{12}\mathbf{P}(\mathcal{U}^{***}_{k_1,k_2}|Y_1, Y_2)\big) = \tilde{p}(k_1+1)\tilde{p}(k_2+1)$$

completes the proof of (11).

Proof of Remark 2. Let us prove (12). The proof is standard and simple. We present it here for reader's convenience. Let (v_1', v_2') denote an ordered pair of distinct vertices drawn uniformly at random and let $\mathbf{P}'$ and $\mathbf{E}'$ denote the conditional probability and expectation given all the random variables considered, but (v_1', v_2'). We have

$$\mathbf{P}(d(v_1) = r|v_1 \sim v_2) = \frac{\mathbf{P}(d(v_1) = r, v_1 \sim v_2)}{\mathbf{P}(v_1 \sim v_2)} = \frac{\mathbf{P}(d(v_1') = r, v_1' \sim v_2')}{\mathbf{P}(v_1 \sim v_2)}. \tag{29}$$

The denominator is evaluated in the Proof of Theorem 3: We have

$$\mathbf{P}(\{v_1 \sim v_2\}) = n^{-1}a_2b_1^2 + o(n^{-1}) = n^{-1}\mathbf{E}\Lambda_3 + o(n^{-1}).$$

In the last step we used the simple identities $\mathbf{E}\Lambda_3 = \mathbf{E}\lambda_3 = a_2b_1^2$. In order to evaluate the numerator we combine identities

$$\mathbf{P}'(d(v_1') = r, v_1' \sim v_2') = \mathbf{P}'(v_1' \sim v_2'|d(v_1') = r)\mathbf{P}'(d(v_1') = r)$$

$$= \frac{r}{n-1}\mathbf{P}'(d(v_1') = r),$$

$$\mathbf{P}(d(v_1') = r, v_1' \sim v_2') = \mathbf{E}\big(\mathbf{P}'(d(v_1') = r, v_1' \sim v_2')\big).$$

We obtain $\mathbf{P}(d(v_1') = r, v_1' \sim v_2') = (r/(n-1))\mathbf{P}(d(v_1) = r)$. Hence, by (29),

$$\mathbf{P}(d(v_1) = r|v_1 \sim v_2) = r\mathbf{P}(d(v_1) = r)(\mathbf{E}\Lambda_3)^{-1} + o(1).$$

Now, the statement (ii) of Theorem 1 completes the proof of (12).

Acknowledgements. The present study was motivated by a question of Konstantin Avratchenkov about the degree-degree distribution in a random intersection graph. A discussion with Konstantin Avratchenkov and Jerzy Jaworski about the influence of the clustering property on the degree-degree distribution was the starting point of this research.

References

1. Aleškevičienė, A., Leipus, R., Šiaulys, J.: Tail behavior of random sums under consistent variation with applications to the compound renewal risk model. Extremes **11**, 261–279 (2008)

2. Avratchenkov, K., Markovich, N.M., Sreedharan, J.K.: Distribution and dependence of extremes in network sampling processes. Research report no. 8578. INRIA, Sophia Antipolis (2014)

3. Bloznelis, M.: Degree-degree distribution in a power law random intersection graph with clustering. Technical report (2014). arXiv:1411.6247

4. Bloznelis, M., Damarackas, J.: Degree distribution of an inhomogeneous random intersection graph. Electron. J. Comb. **20**(3), 3 (2013)

5. Bloznelis, M., Godehardt, E., Jaworski, J., Kurauskas, V., Rybarczyk, K.: Recent progress in complex network analysis - properties of random intersection graphs. In: Lausen, B., Krolak-Schwerdt, S., Böhmer, M. (eds.) Data Science, Learning by Latent Structures, and Knowledge Discovery, pp. 69–78. Springer, Heidelberg (2015)

6. Bloznelis, M., Jaworski, J., Kurauskas, V.: Assortativity and clustering coefficient of sparse random intersection graphs. Electron. J. Probab. **18**(38) (2013)

7. Bloznelis, M., Kurauskas, K.: Clustering function: another view on clustering coefficient. J. Complex Netw. **20**, 1–26 (2015). doi:10.1093/comnet/cnv010

8. Boguna, M., Pastor-Satorras, R., Vespignani, A.: Epidemic spreading in complex networks with degree correlations. Stat. Mech. Complex. Netw. Lect. Notes Phys. **625**, 127–147 (2003)

9. Foss, S., Korshunov, D., Zachary, S.: An Introduction to Heavy-Tailed and Subexponential Distributions, 2nd edn. Springer, New York (2013)

10. Grechnikov, E.A.: The degree distribution and the number of edges between vertices of given degrees in the Buckley-Osthus model of a random web graph. Internet Math. **8**, 257–287 (2012)

11. van der Hofstad, R., Litvak, N.: Degree-degree dependencies in random graphs with heavy-tailed degrees. Internet Math. **10**, 287–334 (2014)

12. Karoński, M., Scheinerman, E.R., Singer-Cohen, K.B.: On random intersection graphs: the subgraph problem. Comb. Probab. Comput. **8**, 131–159 (1999)

13. Muchnik, L., Pei, S., Parra, L.C., Reis, S.D.S., Andrade Jr., J.S., Havlin, S., Makse, H.A.: Origins of power-law degree distribution in the heterogeneity of human activity in social networks. Sci. Rep. **3**, 1783 (2013)

14. Newman, M.E.J.: Assortative mixing in networks. Phys. Rev. Lett. **89**(20), 208701 (2002)

15. Newman, M.E.J.: Mixing patterns in networks. Phys. Rev. E. **67**(2), 026126 (2003)

16. Newman, M.E.J., Strogatz, S.H., Watts, D.J.: Random graphs with arbitrary degree distributions and their applications. Phys. Rev. E. **64**, 026118 (2001)

Upper Bounds for Number of Removed Edges in the Erased Configuration Model

Pim van der Hoorn$^{(\boxtimes)}$ and Nelly Litvak

Department of Electrical Engineering, Mathematics and Computer Science,
University of Twente, Enschede, The Netherlands
w.l.f.vanderhoorn@utwente.nl

Abstract. Models for generating simple graphs are important in the study of real-world complex networks. A well established example of such a model is the erased configuration model, where each node receives a number of half-edges that are connected to half-edges of other nodes at random, and then self-loops are removed and multiple edges are concatenated to make the graph simple. Although asymptotic results for many properties of this model, such as the limiting degree distribution, are known, the exact speed of convergence in terms of the graph sizes remains an open question. We provide a first answer by analyzing the size dependence of the average number of removed edges in the erased configuration model. By combining known upper bounds with a Tauberian Theorem we obtain upper bounds for the number of removed edges, in terms of the size of the graph. Remarkably, when the degree distribution follows a power-law, we observe three scaling regimes, depending on the power law exponent. Our results provide a strong theoretical basis for evaluating finite-size effects in networks.

1 Introduction

The use of complex networks to model large systems has proven to be a powerful tool in recent years. Mathematical and empirical analysis of structural properties of such networks, such as graph distances, centralities, and degree-degree correlations, have received vast attention in recent literature. A common approach for understanding these properties on real-world networks, is to compare them to those of other networks which have the same basic characteristics as the network under consideration, for instance the distribution of the degrees. Such test networks are usually created using random graph models. An important property of real-world networks is that they are usually simple, i.e. there is at most one edge between any two nodes and there are no self-loops. Hence, random graph models that produce simple graphs are of primary interest from the application point of view.

One well established model for generating a graph with given degree distribution is the configuration model [5,19,21], which has been studied extensively in the literature [6,12,15,16]. In this model, each node first receives a certain number of half-edges, or stubs, and then the stubs are connected to each other

© Springer International Publishing Switzerland 2015
D.F. Gleich et al. (Eds.): WAW 2015, LNCS 9479, pp. 54–65, 2015.
DOI: 10.1007/978-3-319-26784-5_5

at random. Obviously, multiple edges and self-loops may appear during the random wiring process. It is well-known that when the degree distribution has finite variance, the graph will be simple with positive probability, so a simple graph can be obtained by repeatedly applying the model until the resulting graph is simple. However, when the variance of the degrees is infinite the resulting graph will, with high probability, not be simple. To remedy this, one can remove self-loops and concatenate the multiple edges to make the graph simple. This version is know as the erased configuration model. Although removal of edges impacts the degree distribution, it has been shown that asymptotically the degree distribution is unchanged. For a thorough systematic treatment of these results we refer the reader to [12].

An important feature of the configuration model is that, conditioned on the graph being simple, it samples a graph uniformly from among all simple graphs with the specified degree distribution. This, in combination with the neutral wiring in the configuration model, makes it a crucial model for studying the effects of degree distributions on the structural properties of the networks, such as, for instance, graph distances [10,13,14,20] and epidemic spread [1,11,17].

We note that there are several different methods for generating simple graphs, sampled uniformly from the set of all simple graphs with a given degree sequence. A large class of such models use Markov-Chain Monte Carlo methods for sampling graphs uniformly from among all graphs with a given set of constraints, such as the degree sequence. These algorithms use so-called edge swap or switching steps, [2,18,23], each time a pair of edges is sampled and swapped, if this is allowed. The main problem with this method are the limited theoretical results on the mixing times, in [7] mixing times are analyzed, but only for regular graphs. Other methods are, for instance, the sequential algorithms proposed in [4,8] which have complexity $O(EN^2)$ and $O(EN)$, respectively, where N is the size of the graph and E denotes the number of edges. The erased configuration model however,is well studied, with strong theoretical results and is easy to implement.

In a recent study [22], authors compare several methods, including the above mentioned Markov-Chain Monte Carlo methods, for creating test graphs for the analysis of structural properties of networks. The authors found that the number of removed edges did not impact the degree sequence in any significant way. However, several other measures on the graph, for instance average neighbor degree, did seem to be altered by the removal of self-loops and double-edges. This emphasizes the fact that asymptotic results alone are not sufficient. The analysis of networks requires a more detailed understanding of finite-size effects in simple random graphs. In particular, it is important to obtain a more precise characterization for dependence of the number of erased edges on the graph size, and their impact on other characteristics of the graph.

In our recent work [16] we analyzed the average number of removed edges in order to evaluate the degree-degree correlations in the directed version of the erased configuration model. We used insights obtained from several limit theorems to derive the scaling in terms of the graph size. Here we make these rigorous by proving three upper bounds for the average number of removed edges in the undirected erased configuration model with regularly varying degree distribution. We start in Sect. 2 with the formal description of the model. Our main result is stated in Sect. 3 and the proofs are provided in Sect. 4.

2 Erased Configuration Model

The Erased Configuration Model (ECM) is an alteration of the Configuration Model (CM), which is a random graph model for generating graphs of size n with either prescribed degree sequence or degree distribution. Given a degree sequence $\mathbf{D}_n$ such that $\sum_{i=1}^{n} D_i$ is even, the degrees of each node are represented as stubs and the graph is constructed by randomly pairing stubs to form edges. This will create a graph with the given degree sequence.

In another version of the model, degrees are sampled independently from a given distribution, an additional stub is added to the last node if the sum of degrees is odd, and the stubs are connected as in the case with given degrees. The empirical degree distribution of the resulting graph will then converge to the distribution from which the degrees were sampled as the graph size goes to infinity, see for instance [12].

When the degree distribution has finite variance, the probability of creating a simple graph with the CM is bounded away from zero. Hence, by repeating the model, one will obtain a simple graph after a finite number of attempts. This construction is called the Repeated Configuration Model (RCM). It has been shown that the RCM samples graphs uniformly from among all simple graphs with the given degree distribution, see Proposition 7.13 in [12].

When the degrees have infinite variance, the probability of generating a simple graph with the CM converges to zero as the graph size increases. In this case the ECM can be used, where after all stubs are paired, multiple edges are merged and self-loops are removed. This model is computationally far less expensive than the RCM since the pairing only needs to be done once while in the other case the number of attempts increases as the variance of the degree distribution grows. The trade-off is that the ECM removes edges, altering the degree sequence and hence the empirical degree distribution. Nevertheless it was proven, see [12], that the empirical degree distribution for the ECM still converges to the original one as $n \to \infty$.

For our analysis we shall consider graphs of size n generated by the ECM, where the degrees are sampled at random from a regularly varying distribution. We recall that X is said to have a regularly varying distribution with finite mean if

$$\mathbb{P}\left(X > k\right) = L(k)k^{-\gamma} \quad \text{with } \gamma > 1, \tag{1}$$

where L is a slowly varying function, i.e. $\lim_{x \to \infty} L(tx)/L(x) = 1$ for all $t > 0$. The parameter γ is called the exponent of the distribution.

For $n \in \mathbb{N}$ we consider the degree sequence $\mathbf{D}_n$ as a sequence of i.i.d. samples from distribution (1), let $\mu = \mathbb{E}[D]$ and denote by $L_n = \sum_{i=1}^{n} D_i$ the sum of the degrees. Formally we need L_n to be even in order to have a graphical sequence, in which case $L_n/2$ is the number of edges. This can be achieved by increasing the degree of the last node, D_n, by one if the sum is odd. This alteration adds a term uniformly bounded by one which does not influence the analysis. Therefore we can omit this and treat the degree sequence $\mathbf{D}_n$ as an i.i.d. sequence.

For the analysis we denote by E_{ij} the number of edges between two nodes, $1 \leq i, j \leq n$, created by the CM and by E_{ij}^c the number of edges between the two nodes that where removed by the ECM. Furthermore, we let $\mathbb{P}_n$ and $\mathbb{E}_n$ be, respectively, the probability and expectation conditioned on the degree sequence $\mathbf{D}_n$.

3   Main Result

The main result of this paper is concerned with the scaling of the average number of erased edges in the ECM. It was proven in [15] that

$$\frac{1}{L_n} \sum_{i,j} \mathbb{E}_n \left[E_{ij}^c \right] \xrightarrow{\mathbb{P}} 0 \quad \text{as } n \to \infty, \tag{2}$$

where $\xrightarrow{\mathbb{P}}$ denotes convergence in probability. This result states that the average number of removed edges converges to zero as the graph size grows, which is in agreement with the convergence in probability of the empirical degree distribution to the original one. However, until now there have not been any results on the speed of convergence in (2). In this section we will state our result, which establishes upper bounds on the scaling of the average number of erased edges.

To make our statement rigorous we first need to define what we mean by scaling for a random variable.

Definition 1. *Let* $(X_n)_{n \in \mathbb{N}}$ *be sequences of random variables and let* $\rho \in \mathbb{R}$. *Then we define*

$$X_n = O_{\mathbb{P}} \left(n^\rho \right) \iff \text{for all } \varepsilon > 0 \quad n^{-\rho-\varepsilon} X_n \xrightarrow{\mathbb{P}} 0, \quad \text{as } n \to \infty.$$

We are now ready to state the main result on the scaling of the average number of erased edges in the ECM.

Theorem 1. *Let* G_n *be a graph generated by the ECM with degree distribution* (1), *let* L_n *be the sum of the degrees and denote by* E_{ij}^c *the number of removed edges from* i *to* j. *Then*

$$\frac{1}{L_n} \sum_{i,j=1}^n \mathbb{E}_n \left[E_{ij}^c \right] = \begin{cases} O_{\mathbb{P}} \left(n^{\frac{1}{\gamma}-1} \right) & \text{if } 1 < \gamma \leq \frac{3}{2}, \\ O_{\mathbb{P}} \left(n^{\frac{4}{\gamma}-3} \right) & \text{if } \frac{3}{2} < \gamma \leq 2, \\ O_{\mathbb{P}} \left(n^{-1} \right) & \text{if } \gamma > 2. \end{cases} \tag{3}$$

The proof of Theorem 1 is given in the next section. The strategy of the proof is to establish two upper bounds for $\sum_{i,j=1}^n \mathbb{E}_n \left[E_{ij}^c \right]/L_n$ for the case $1 < \gamma \leq 2$, each of which scales as one of the first two terms from (3). Then it remains to

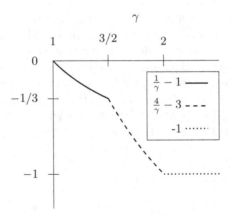

Fig. 1. The scaling exponent of the average number of erased edges, as a function of γ.

observe that the term $n^{1/\gamma-1}$ dominates $n^{4/\gamma-3}$ when $1 < \gamma \leq 3/2$ while the latter one dominates when $3/2 < \gamma < 2$. In addition, we prove the n^{-1} scaling for $\gamma > 2$.

Theorem 1 gives several insights into the behavior of the ECM. First, consider the case when the degrees have finite variance ($\gamma > 2$). Equation (3) tells us that in that case the ECM will erase only a finite, in n, number of edges. For large n, this alters the degree sequence in a negligible way. We then gain the advantage that we need to perform the random wiring only once. In contrast, the RCM requires multiple attempts before a simple graph is produced. This will be a problem, especially as γ approaches 2.

An even more interesting phenomenon established by Theorem 1 is the remarkable change in the scaling at $\gamma = 3/2$. Figure 1 shows the exponent in the scaling term in (3), as a function of γ. Notice that for small values of γ, the fraction of erased edges decreases quite slowly with n. For example, when $\gamma = 1.1$ and $n = 10^6$ then $n^{1/\gamma} \approx 284803$. Hence, a significant fraction of edges will be removed, so we can expect notable finite size effects even in large networks. However, when $\gamma \geq 3/2$ the finite size effects are already very small and decrease more rapidly with γ.

It will be seen from the proofs in the next section that the upper bounds for $\gamma > 3/2$ in Theorem 1 follow readily from the literature. Our main contribution is in the upper bound for $1 < \gamma < 3/2$, which corresponds to many real-world networks. The proof uses a Central Limit Theorem and a Tauberian Theorem for regularly varying random variables. Note that when $1 < \gamma < 3/2$ the upper bound $n^{4/\gamma-3}$ is not at all tight and even increasing in n for $\gamma < 4/3$.

4   Upper Bounds for Erased Edges

Throughout this section we will use the Central Limit Theorem for regularly varying random variables also called the Stable Law CLT, see [24] Theorem

4.5.1. We summarize it below, letting $\xrightarrow{d}$ denote convergence in distribution, in the setting of non-negative regularly varying random variables.

Theorem 2 (Stable Law CLT [24]). *Let $\{X_i : i \geq 1\}$ be an i.i.d. sequence of non-negative random variables with distribution (1) and $0 < \gamma < 2$. Then there exists a slowly varying function L_0, different from L, such that*

$$\frac{\sum_{i=1}^n D_i - m_n}{L_0(n)n^{\frac{1}{\gamma}}} \xrightarrow{d} S_\gamma,$$

where S_γ is a stable random variable and

$$m_n = \begin{cases} 0 & \text{if } 0 < \gamma < 1 \\ n^2 \mathbb{E}\left[\sin\left(\frac{X}{L_0(n)n}\right)\right] & \text{if } \gamma = 1 \\ n\mathbb{E}[X] & \text{if } 1 < \gamma < 2. \end{cases}$$

From Theorem 2 we can infer several scaling results using the following observation: By Slutsky's Theorem it follows that

$$\frac{\sum_{i=1}^n X_i - m_n}{L_0(n)n^{\frac{1}{\gamma}}} \xrightarrow{d} S_\gamma \quad \text{as } n \to \infty$$

implies that for any $\varepsilon > 0$,

$$n^{-\varepsilon}\frac{\sum_{i=1}^n X_i - m_n}{L_0(n)n^{\frac{1}{\gamma}}} \xrightarrow{\mathbb{P}} 0 \quad \text{as } n \to \infty.$$

Hence $|\sum_{i=1}^n X_i - m_n| = O_{\mathbb{P}}\left(L_0(n)n^{1/\gamma}\right)$ and therefore, by Potter's Theorem, it follows that $|\sum_{i=1}^n X_i - m_n| = O_{\mathbb{P}}\left(n^{1/\gamma}\right)$. Finally, we remark that if D has distribution (1) with $1 < \gamma \leq 2$, then D^2 has distribution (1) with exponent $1/2 < \gamma/2 \leq 1$. Summarizing, we have the following.

Corollary 1. *Let G_n be a graph generated by the ECM with degree distribution (1) and $1 < \gamma \leq 2$, then*

$$L_n = O_{\mathbb{P}}(n), \quad \left|\sum_{i=1}^n D_i - \mu n\right| = O_{\mathbb{P}}\left(n^{\frac{1}{\gamma}}\right) \text{ and } \sum_{i=1}^n D_i^2 = O_{\mathbb{P}}\left(n^{\frac{2}{\gamma}}\right)$$

The third equation also holds for $\gamma = 2$ since

$$\sum_{i=1}^n D_i^2 = \left(\sum_{i=1}^n D_i^2 - L_0(n)n^2\mathbb{E}\left[\sin\left(\frac{D}{nL_0(n)}\right)\right]\right) + L_0(n)n^2\mathbb{E}\left[\sin\left(\frac{D}{n^1 L_0(n)}\right)\right]$$

$$\leq \left(\sum_{i=1}^n D_i^2 - n^2 L_0(n)\mathbb{E}\left[\sin\left(\frac{D}{n^1 L_0(n)}\right)\right]\right) + n\mu$$

$$= O_{\mathbb{P}}(L_0(n)n) + n\mu = O_{\mathbb{P}}(n).$$

4.1   The Upper Bounds $O_{\mathbb{P}}\left(n^{\frac{4}{\gamma}-3}\right)$ and $O_{\mathbb{P}}\left(n^{-1}\right)$

For the proof of the upper bounds we will use the following proposition.

Proposition 1 (Proposition 7.10 [12]). *Let G_n be a graph generated by the CM and denote by S_n and M_n, respectively, the number of self-loops and multiple edges. Then*

$$\mathbb{E}_n\left[S_n\right] \leq \sum_{i=1}^{n} \frac{D_i^2}{L_n} \quad and \quad \mathbb{E}_n\left[M_n\right] \leq 2\left(\sum_{i=1}^{n} \frac{D_i^2}{L_n}\right)^2.$$

Lemma 1. *Let G_n be a graph generated by the ECM with degree distribution (1), then*

$$\frac{1}{L_n} \sum_{i,j=1}^{n} \mathbb{E}_n\left[E_{ij}^c\right] = \begin{cases} O_{\mathbb{P}}\left(n^{\frac{4}{\gamma}-3}\right) & \text{if } 1 < \gamma \leq 2, \\ O_{\mathbb{P}}\left(n^{-1}\right) & \text{if } \gamma > 2. \end{cases} \tag{4}$$

Proof. We start by observing that

$$\sum_{i,j=1}^{n} E_{ij}^c = S_n + M_n,$$

and hence it follows from Proposition 1 that

$$\sum_{i,j=1}^{n} \mathbb{E}_n\left[E_{ij}^c\right] \leq \sum_{i=1}^{n} \frac{D_i^2}{L_n} + 2\left(\sum_{i=1}^{n} \frac{D_i^2}{L_n}\right)^2.$$

First suppose that $1 < \gamma \leq 2$. Then, by Corollary 1 and the continuous mapping theorem it follows that

$$\frac{1}{L_n} \sum_{i,j=1}^{n} \mathbb{E}_n\left[E_{ij}^c\right] \leq \frac{1}{L_n^2} \sum_{i=1}^{n} D_i^2 + 2\frac{1}{L_n^3}\left(\sum_{i=1}^{n} D_i^2\right)^2 = O_{\mathbb{P}}\left(n^{\frac{4}{\gamma}-3}\right).$$

Now suppose that $\gamma > 2$. Then D_i^2 has finite mean, say ν, and therefore, by Theorem 2,

$$\frac{1}{L_n^2} \sum_{i=1}^{n} D_i^2 \leq \frac{1}{L_n^2}\left|\sum_{i=1}^{n} D_i^2 - n\nu\right| + \frac{n\nu}{L_n^2} = O_{\mathbb{P}}\left(n^{\frac{2}{\gamma}-2} + n^{-1}\right) = O_{\mathbb{P}}\left(n^{-1}\right),$$

where the last step follows since $2/\gamma - 2 < -1$ when $\gamma > 2$. Since this is the main term the result follows.   □

Lemma 1 provides the last two upper bounds from Theorem 1. However, as we mentioned before, the bound $O_{\mathbb{P}}\left(n^{4/\gamma-3}\right)$ is not tight over the whole range $1 < \gamma \leq 2$ since for $1 < \gamma < 4/3$ we have $4/\gamma - 3 > 0$, and hence the upper bound diverges as $n \to \infty$ which is in disagreement with (2). Therefore, there must exist another upper bound on the average erased number of edges, which goes to zero as $n \to \infty$ for all $\gamma > 1$. This new bound does not follow readily from the literature. Below we establish such upper bound and explain the essential new ingredients needed for its proof.

4.2 The Upper Bound $O_{\mathbb{P}}\left(n^{\frac{1}{\gamma}-1}\right)$

We first observe that the number of erased edges between nodes i and j equals the total number of edges between the nodes minus one, if there is more than one edge. This gives,

$$
\frac{1}{L_n} \sum_{i,j=1}^{n} \mathbb{E}_n\left[E_{ij}^c\right] = \frac{1}{L_n} \sum_{i,j=1}^{n} \mathbb{E}_n\left[E_{ij} - \mathbb{1}_{\{E_{ij}>0\}}\right]
$$

$$
= \frac{1}{L_n} \sum_{i,j=1}^{n} \mathbb{E}_n\left[E_{ij}\right] - \frac{1}{L_n} \sum_{i,j=1}^{n} \mathbb{E}_n\left[1 - \mathbb{1}_{\{E_{ij}=0\}}\right]
$$

$$
= 1 - \frac{n^2}{L_n} + \frac{1}{L_n} \sum_{i,j=1}^{n} \mathbb{P}_n\left(E_{ij}=0\right). \tag{5}
$$

We can get an upper bound for $\mathbb{P}_n\left(E_{ij}=0\right)$ from the analysis performed in [13], Sect. 4. Since the probability of no edges between i and j equals the probability that none of the D_i stubs connects to one of the D_j stubs, it follows from Eq. (4.9) in [13] that

$$
\mathbb{P}_n\left(E_{ij}=0\right) \leq \prod_{k=0}^{D_i-1}\left(1 - \frac{D_j}{L_n - 2D_i - 1}\right) + \frac{D_i^2 D_j}{(L_n - 2D_i)^2}. \tag{6}
$$

The product term in (6) can be upper bounded by $\exp\{-D_i D_j / E_n\}$. For the second term we use that

$$
\frac{1}{L_n} \sum_{i,j=1}^{n} \frac{D_i^2 D_j}{(L_n - 2D_i)^2} = \frac{1}{L_n^2} \sum_{i=1}^{n} D_i^2 \left(\frac{1}{1 - \frac{2D_i}{L_n}}\right)^2 \left(\frac{1}{L_n} \sum_{j=1}^{n} D_j\right)
$$

$$
\leq \frac{1}{L_n^2} \sum_{i=1}^{n} D_i^2 = O_{\mathbb{P}}\left(n^{\frac{2}{\gamma}-2}\right).
$$

Putting everything together we obtain

$$
\frac{1}{L_n} \sum_{i,j=1}^{n} \mathbb{P}_n\left(E_{ij}=0\right) \leq \sum_{i,j=1}^{n} \exp\left\{-\frac{D_i^+ D_j^-}{L_n}\right\} + O_{\mathbb{P}}\left(n^{\frac{2}{\gamma}-2}\right). \tag{7}
$$

We will use (7) to upper bound (5). In order to obtain the desired result we will employ a Tauberian Theorem for regularly varying random variables, which we summarize first. We write $a \sim b$ to denote that a/b goes to one in a corresponding limit.

Theorem 3 (Tauberian Theorem, [3]). *Let X be a non-negative random variable with only finite mean. Then, for $1 < \gamma < 2$, the following are equivalent,*

(i) $\mathbb{P}\left(X > t\right) \sim L(t)t^{-\gamma}$ as $t \to \infty$,

(ii) $\dfrac{\mathbb{E}[X]}{t} - 1 + \exp\left\{-\dfrac{X}{t}\right\} \sim L\left(\dfrac{1}{t}\right) t^{-\gamma} \quad$ as $t \to \infty$.

We will first explain the idea behind the proof of the $O_{\mathbb{P}}\left(n^{1/\gamma-1}\right)$ bound. If we insert (7) into (5) we get

$$\frac{1}{L_n} \sum_{i,j=1}^{n} \mathbb{E}_n\left[E_{ij}^c\right] \leq 1 - \frac{n^2}{L_n} + \frac{1}{L_n} \sum_{i,j=1}^{n} \exp\left\{-\frac{D_i^+ D_j^-}{L_n}\right\} + O_{\mathbb{P}}\left(n^{\frac{2}{\gamma}-2}\right). \quad (8)$$

The terms on the right side can be rewritten to obtian an expression that resembles an empirical version of the left hand side of part (ii) from Theorem 3, with $t = L_n$ and $X = D_1 D_2$. Thus, the scaling of the average number of erased edges will be determined by the scaling that follows from the Tauberian Theorem and the Stable Law CLT.

Proposition 2. *Let G_n be a graph generated by the ECM with degree distribution (1) and $1 < \gamma < 2$. Then*

$$\frac{1}{L_n} \sum_{i,j=1}^{n} \mathbb{E}_n\left[E_{ij}^c\right] = O_{\mathbb{P}}\left(n^{\frac{1}{\gamma}-1}\right). \quad (9)$$

Proof. We start with Eq. (8). Since the correction term here is of lower order, by extracting a factor n^2/L_n from the other terms and using that $L_n = \sum_{i=1}^{n} D_i$, it suffices to show that

$$\frac{n^2}{L_n}\left(\frac{1}{n^2}\sum_{i,j=1}^{n}\frac{D_i D_j}{L_n} - 1 + \frac{1}{n^2}\sum_{i,j=1}^{n}\exp\left\{-\frac{D_i D_j}{L_n}\right\}\right) = O_{\mathbb{P}}\left(n^{\frac{1}{\gamma}-1}\right). \quad (10)$$

We first consider the term inside the brackets in the left hand side of (10).

$$\left|\frac{1}{n^2}\sum_{i,j=1}^{n}\frac{D_i D_j}{L_n} - 1 + \frac{1}{n^2}\sum_{i,j=1}^{n}\exp\left\{-\frac{D_i D_j}{L_n}\right\}\right|$$

$$\leq \frac{1}{n^2}\left|\frac{1}{L_n} - \frac{1}{\mu n}\right|\sum_{i,j=1}^{n} D_i D_j \quad (11)$$

$$+ \frac{1}{n^2}\sum_{i,j=1}^{n}\left|\exp\left\{-\frac{D_i D_j}{L_n}\right\} - \exp\left\{-\frac{D_i D_j}{\mu n}\right\}\right| \quad (12)$$

$$+ \left|\frac{1}{n^2}\sum_{i,j=1}^{n}\left(\frac{D_i D_j}{\mu n} - 1 + \exp\left\{-\frac{D_i D_j}{\mu n}\right\}\right)\right| \quad (13)$$

Since

$$\frac{1}{n^2}\sum_{i,j=1}^{n}\left|\exp\left\{-\frac{D_i D_j}{L_n}\right\} - \exp\left\{-\frac{D_i D_j}{\mu n}\right\}\right| \leq \frac{1}{n^2}\left|\frac{1}{L_n} - \frac{1}{\mu n}\right|\sum_{i,j=1}^{n} D_i D_j,$$

it follows from Corollary 1 that both (11) and (12) are $O_{\mathbb{P}}\left(n^{\frac{1}{\gamma}-2}\right)$. Next, observe that the function $e^{-x} - 1 + x$ is positive which implies, by Markov's inequality, that (13) scales as its average

$$\frac{\mathbb{E}\left[D_1 D_2\right]}{\mu n} - 1 + \mathbb{E}\left[\exp\left\{-\frac{D_1 D_2}{\mu n}\right\}\right]. \tag{14}$$

where D_1 and D_2 are two independent random variables with distribution (1) and $1 < \gamma < 2$, so that the product $D_1 D_2$ again has distribution (1) with the same exponent, see for instance the Corollary after Theorem 3 in [9]. Now we use Theorem 3 to find that (14), and hence (13) are $O_{\mathbb{P}}\left(n^{-\gamma}\right)$. Finally, the term outside of the brackets in (10) is $O_{\mathbb{P}}\left(n\right)$ and since $1 - \gamma < \frac{1}{\gamma} - 1$ for all $\gamma > 1$, the result follows. □

5   Discussion

The configuration model is one of the most important random graph models developed so far for constructing test graphs, used in the study of structural properties of, and processes on, real-world networks. The model is of course most true to reality when it produces a simple graph. Because this will happen with vanishing probability for most networks, since these have infinite degree variance, the ECM can be seen as the primary model for a neutrally wired simple graph with scale-free degrees. The fact that the fraction of erased edges is vanishing, suffices for obtaining asymptotic structural properties and asymptotic behavior of network processes in the ECM. However, real-world networks are finite, albeit very large. Therefore, it is important to understand and quantify how the properties and processes in a finite network are affected by the fact that the graph is simple.

This paper presents the first step in this direction by providing probabilistic upper bounds for the number of the erased edges in the undirected ECM. This second order analysis shows that the average number of erased edges by the ECM decays as n^{-1} when the variance of the degrees is finite. Since the ECM is computationally less expensive then the RCM and other sequential algorithms, this is a strong argument for using the ECM as a standard model for generating test graphs with given degree distribution. Especially since, in contrast to Markov-Chain Monte Carlo methods using edge swap mechanics, it is theoretically well analyzed. We also uncover a new transition in the scaling of the average number of erased edges for regularly varying degree distributions with only finite mean, in terms of the exponent of the degree distribution.

Based on the empirical results found by us in [16], we conjecture that the bounds we obtained are tight, up to some slowly varying functions. Therefore, as a next step one could try to prove Central Limit Theorems for the number of erased edges, using the bounds from Theorem 1 as the correct scaling factors. These tools would make it possible to perform statistical analysis of properties on networks, using the ECM as a model for generating test graphs.

References

1. Andersson, H.: Limit theorems for a random graph epidemic model. Ann. Appl. Probab. **8**, 1331–1349 (1998)
2. Artzy-Randrup, Y., Stone, L.: Generating uniformly distributed random networks. Phys. Rev. E **72**(5), 056708 (2005)
3. Bingham, N.H., Doney, R.A.: Asymptotic properties of supercritical branching processes i: the galton-watson process. Adv. Appl. Probab. **6**, 711–731 (1974)
4. Blitzstein, J., Diaconis, P.: A sequential importance sampling algorithm for generating random graphs with prescribed degrees. Internet Math. **6**(4), 489–522 (2011)
5. Bollobás, B.: A probabilistic proof of an asymptotic formula for the number of labelled regular graphs. Eur. J. Comb. **1**(4), 311–316 (1980). http://www.sciencedirect.com/science/article/pii/S0195669880800308
6. Britton, T., Deijfen, M., Martin-Löf, A.: Generating simple random graphs with prescribed degree distribution. J. Stat. Phys. **124**(6), 1377–1397 (2006)
7. Cooper, C., Dyer, M., Greenhill, C.: Sampling regular graphs and a peer-to-peer network. Comb. Probab. Comput. **16**(04), 557–593 (2007)
8. Del Genio, C.I., Kim, H., Toroczkai, Z., Bassler, K.E.: Efficient and exact sampling of simple graphs with given arbitrary degree sequence. PloS One **5**(4), e10012 (2010)
9. Embrechts, P., Goldie, C.M.: On closure and factorization properties of subexponential and related distributions. J. Aust. Math. Soc. (Series A) **29**(02), 243–256 (1980)
10. van den Esker, H., van der Hofstad, R., Hooghiemstra, G., Znamenski, D.: Distances in random graphs with infinite mean degrees. Extremes **8**(3), 111–141 (2005)
11. Ferreira, S.C., Castellano, C., Pastor-Satorras, R.: Epidemic thresholds of the susceptible-infected-susceptible model on networks: a comparison of numerical and theoretical results. Phys. Rev. E **86**(4), 041125 (2012)
12. van der Hofstad, R.: Random graphs and complex networks. Unpublished manuscript (2007). http://www.win.tue.nl/rhofstad/NotesRGCN.pdf
13. van der Hofstad, R., Hooghiemstra, G., Van Mieghem, P.: Distances in random graphs with finite variance degrees. Random Struct. Algorithms **27**(1), 76–123 (2005)
14. van der Hofstad, R., Hooghiemstra, G., Znamenski, D.: Distances in random graphs with finite mean and infinite variance degrees. Eurandom (2005)
15. van der Hoorn, P., Litvak, N.: Convergence of rank based degree-degree correlations in random directed networks. Moscow J. Comb. Number Theor. **4**(4), 45–83 (2014). http://mjcnt.phystech.edu/en/article.php?id=92
16. van der Hoorn, P., Litvak, N.: Phase transitions for scaling of structural correlations in directed networks (2015). arXiv preprint arXiv:1504.01535
17. Lee, H.K., Shim, P.S., Noh, J.D.: Epidemic threshold of the susceptible-infected-susceptible model on complex networks. Phys. Rev. E **87**(6), 062812 (2013)
18. Maslov, S., Sneppen, K.: Specificity and stability in topology of protein networks. Science **296**(5569), 910–913 (2002)
19. Molloy, M., Reed, B.: A critical point for random graphs with a given degree sequence. Random Struct. Algorithms **6**(2–3), 161–180 (1995). http://onlinelibrary.wiley.com/doi/10.1002/rsa.3240060204/full
20. Molloy, M., Reed, B.: The size of the giant component of a random graph with a given degree sequence. Comb. Probab. Comput. **7**(03), 295–305 (1998)

21. Newman, M.E., Strogatz, S.H., Watts, D.J.: Random graphs with arbitrary degree distributions and their applications. Phys. Rev. E **64**(2), 026118 (2001). http://journals.aps.org/pre/abstract/10.1103/PhysRevE.64.026118

22. Schlauch, W.E., Horvát, E.Á., Zweig, K.A.: Different flavors of randomness: comparing random graph models with fixed degree sequences. Soc. Netw. Anal. Min. **5**(1), 1–14 (2015)

23. Tabourier, L., Roth, C., Cointet, J.P.: Generating constrained random graphs using multiple edge switches. J. Exp. Algorithmics (JEA) **16**, 1–7 (2011)

24. Whitt, W.: Stochastic-Process Limits: An Introduction to Stochastic-Process Limits and Their Application to Queues. Springer, New York (2002)

The Impact of Degree Variability on Connectivity Properties of Large Networks

Lasse Leskelä[(✉)] and Hoa Ngo

Department of Mathematics and Systems Analysis,
School of Science, Aalto University, PO Box 11100, 00076 Aalto, Finland
{lasse.leskela,hoa.ngo}@aalto.fi

Abstract. The goal of this work is to study how increased variability in the degree distribution impacts the global connectivity properties of a large network. We approach this question by modeling the network as a uniform random graph with a given degree sequence. We analyze the effect of the degree variability on the approximate size of the largest connected component using stochastic ordering techniques. A counterexample shows that a higher degree variability may lead to a larger connected component, contrary to basic intuition about branching processes. When certain extremal cases are ruled out, the higher degree variability is shown to decrease the limiting approximate size of the largest connected component.

Keywords: Configuration model · Size-biased distribution · Length-biased distribution · Weighted distribution · Convex stochastic order · Stochastic comparisons

1 Introduction

Digital communication networks and online social media have dramatically increased the spread of information in our society. As a result, the global connectivity structure of communication between people appears to be better modeled as a dimension-free unstructured graph instead of a geometrical graph based on a two-dimensional grid, and the spread of messages over an online network can be modeled as an epidemic on a large random graph. When the nodes of the network spread the epidemic independently of each other, the final outcome of the epidemic, or the ultimate set of nodes that receive a message, corresponds to the connected component of the initial root node in a randomly thinned version of the original communication graph called the epidemic generated graph [1]. This is why the sizes of connected components are important in studying information dynamics in unstructured networks.

A characterizing statistical feature of many communication networks is the high variability among node degrees, which is manifested by observed approximate power-law shapes in empirical measurements. The simplest mathematical

H. Ngo—Research supported by the Emil Aaltonen Foundation.

D.F. Gleich et al. (Eds.): WAW 2015, LNCS 9479, pp. 66–77, 2015.
DOI: 10.1007/978-3-319-26784-5_6

model that allows to capture the degree variability is the so-called configuration model which is defined as follows. Fix a set of nodes labeled using $[n] = \{1, 2, \ldots, n\}$ and a sequence of nonnegative integers $d_n = \{d_n(1), \ldots, d_n(n)\}$ such that $\ell_n = \sum_{i=1}^{n} d_n(i)$ is even. Each node i gets assigned $d_n(i)$ half-links, or stubs, and then we select a uniform random matching among the set of all half-links. A matched pair of half-links will form a link, and we denote by $X_{i,j}$ the number of links with one half-link assigned to i and the other half-link assigned to j. The resulting random matrix $(X_{i,j})$ constitutes a random undirected multigraph on the node set $[n]$. This multigraph is called the *configuration model* generated by the degree sequence d_n. The multigraph is called simple if it contains no loops ($X_{i,i} = 0$ for all i) and no parallel links ($X_{i,j} \leq 1$ for all i, j). The distribution of the multigraph conditional on being simple is the same as the distribution of the uniform random graph in the space of graphs on $[n]$ with degree sequence d_n [4, Proposition 7.13].

A tractable mathematical way to analyze large random graphs is to let the size of the graph grow to infinity and approximate the empirical degree distribution of the random graph

$$p_n(k) = \frac{1}{n} \sum_{i=1}^{n} 1(d_n(i) = k)$$

using a limiting probability distribution p on the infinite set of nonnegative integers $\mathbb{Z}_+$. One of the key results in the theory of random graphs is the following, first derived by Molloy and Reed [7,8] and strengthened by Janson and Łuczak [5]. Assume that the collection of degree sequences (d_n) is such that the corresponding empirical degree distributions satisfy

$$p_n(k) \to p(k) \quad \text{for all } k \geq 0,$$
$$\sup_n m_2(p_n) < \infty, \tag{1}$$

and that $p(2) < 1$ and $0 < m_1(p) < \infty$, where $m_i(p) = \sum_k k^i p(k)$ denotes the ith moment p. Then [5, Theorem 2.3, Remark 2.7] the size of the largest connected component $|\mathcal{C}_{\max}|$ in the configuration model multigraph (and in the associated uniform random graph) converges according to

$$n^{-1}|\mathcal{C}_{\max}| \to \zeta_{\mathrm{CM}}(p) \quad \text{(in probability)}, \tag{2}$$

where the constant $\zeta_{\mathrm{CM}}(p) \in [0, 1]$ is uniquely characterized by p and satisfies $\zeta_{\mathrm{CM}}(p) > 0$ if and only if $m_2(p) > 2m_1(p)$. The above fundamental result is important because it has been extended to models of wide generality (e.g. [2]).

Most earlier mathematical studies (and extensions) have focused on establishing the phase transition (showing that there is a critical phenomenon related to whether or not $\zeta_{\mathrm{CM}}(p) > 0$), and studying the behavior of the model near the critical regime. On the other hand, for practical applications it may crucial to be able to predict the size of $\zeta_{\mathrm{CM}}(p)$ based on estimates of the degree distribution p. This paper aims to obtain qualitative insight into this question by studying properties of the functional $p \mapsto \zeta_{\mathrm{CM}}(p)$ in detail analyzing its sensitivity to the variability of p.

2   The Branching Functional of the Configuration Model

2.1   Size Biasing and Downshifting

The configuration model, like many real-world networks, exhibits a size-bias phenomenon in degrees, in that "your friends are likely to have more friends than you do". The *size biasing* of a probability distribution μ on the nonnegative real line $\mathbb{R}_+$ (or a subset thereof) with mean $m_1(\mu) = \int x\mu(dx) \in (0, \infty)$, is the probability distribution μ^* defined by

$$\mu^*(B) = \frac{\int_B x\mu(dx)}{m_1(\mu)}, \quad B \subset \mathbb{R}_+.$$

If X and X^* are random numbers with distributions μ and μ^*, respectively, then

$$\mathbb{E}\,\phi(X^*) = \frac{\mathbb{E}\,\phi(X)X}{\mathbb{E}X} \tag{3}$$

for any real function ϕ such that the above expectations exist. The size biasing of a probability distribution p on the nonnegative integers $\mathbb{Z}_+$ is given by

$$p^*(k) = \frac{kp(k)}{m_1(p)}, \quad k \in \mathbb{Z}_+.$$

Furthermore, the *downshifted size biasing* of p is the probability distribution p° defined by

$$p^\circ(k) = p^*(k+1), \quad k \in \mathbb{Z}_+. \tag{4}$$

If X^* and X° are random integers distributed according to p^* and p°, respectively, then X° and $X^* - 1$ are equal in distribution.

Example 1. The size biasing of the Dirac point mass at x is given by $\delta_x^* = \delta_x$.

Example 2. The size biasing of the Pareto distribution $\mathrm{Par}(\alpha, c)$ on $\mathbb{R}_+$ with shape $\alpha > 1$ and scale $c > 0$ (with density function $f(t) = \alpha c^\alpha t^{-\alpha-1}1(t > c)$) is given by $\mathrm{Par}(\alpha, c)^* = \mathrm{Par}(\alpha - 1, c)$.

Example 3. Denote by $\mathrm{MPoi}(\mu)$ the μ-mixed Poisson distribution on $\mathbb{Z}_+$ with probability mass function

$$p(k) = \int_{\mathbb{R}_+} e^{-\lambda k} \frac{\lambda^k}{k!}\, \mu(d\lambda), \quad k \in \mathbb{Z}_+,$$

where μ is a probability distribution on $\mathbb{R}_+$ with a finite nonzero mean. In this case the downshifted size biasing is given by $\mathrm{MPoi}(\mu)^\circ = \mathrm{MPoi}(\mu^*)$. Especially, by Example 1, $\mathrm{Poi}(x)^\circ = \mathrm{MPoi}(\delta_x)^\circ = \mathrm{Poi}(x)$ for a standard Poisson distribution with mean x, and by Example 2, $\mathrm{MPoi}(\mathrm{Par}(\alpha, c))^\circ = \mathrm{MPoi}(\mathrm{Par}(\alpha - 1, c))$ for a Pareto-mixed Poisson distribution with shape $\alpha > 1$ and scale $c > 0$.

2.2   Branching Functional of the Configuration Model

Given a probability distribution p on $\mathbb{Z}_+$, we denote by

$$\eta(p) = \inf\{s \geq 0 : G_p(s) = s\}$$

the smallest fixed point of the generating function $G_p(s) = \sum_{k \geq 0} s^k p(k)$ in the interval $[0, 1]$. Classical branching process theory (e.g. [3,4]) tells that $\eta(p) \in [0, 1]$ is well defined and equal to the extinction probability of a Galton–Watson process with offspring distribution p. We denote the corresponding survival probability by

$$\zeta(p) = 1 - \eta(p). \tag{5}$$

As a consequence of [5, Theorem 2.3], the limiting maximum component size of a configuration model with limiting degree distribution p corresponds to the survival probability of a two-stage branching process where the root node has offspring distribution p and all other nodes have offspring distribution p° defined by (4). Therefore, the branching functional $p \mapsto \zeta_{CM}(p)$ appearing in (2) can be written as

$$\zeta_{CM}(p) = 1 - G_p(\eta(p^\circ)). \tag{6}$$

A simple closed-form expression for $\zeta_{CM}(p)$ is not readily available due to the implicit definition of $\eta(p^\circ)$. To get a qualitative insight into the behavior of $\zeta_{CM}(p)$ as a functional of p, analytical upper and lower bounds will be valuable tools. The following result provides a first crude upper bound. Similar bounds for standard branching processes have been derived in [10,12].

Proposition 1. *For any probability distribution p on $\mathbb{Z}_+$ with a finite nonzero mean $m_1(p)$,*

$$\zeta_{CM}(p) \leq 1 - p(0) - \frac{p(1)^2}{m_1(p)}. \tag{7}$$

Proof. Let p° be the downshifted size biasing of p defined by (4). Because a branching process with offspring distribution p° goes extinct at the first step with probability $p^\circ(0)$, it follows that

$$\eta(p^\circ) \geq p^\circ(0) = \frac{p(1)}{m_1(p)}.$$

Together with $G_p(s) \geq p(0) + p(1)s$, this shows that

$$G_p(\eta(p^\circ)) \geq p(0) + \frac{p(1)^2}{m_1(p)}.$$

The above inequality substituted into (6) implies (7).

3   Ordering of Branching Processes

3.1   Strong and Convex Stochastic Orders

The upper bound of $\zeta_{\mathrm{CM}}(p)$ obtained in Proposition 1 is rough as it disregards information about the tail characteristics of p. To obtain better estimates, we will develop in this section techniques based on the theory of stochastic orders (see [9] or [11] for comprehensive surveys).

Integral stochastic orderings between probability distributions on $\mathbb{R}$ (or a subset thereof) are defined by requiring

$$\int \phi(x)\mu(dx) \leq \int \phi(x)\nu(dx) \tag{8}$$

to hold for all functions $\phi : \mathbb{R} \to \mathbb{R}$ in a certain class of functions such that both integrals above exist. The *strong stochastic order* is defined by denoting $\mu \leq_{\mathrm{st}} \nu$ if (8) holds for all increasing functions ϕ. The *convex stochastic order* (resp. concave, increasing convex, increasing concave) order is defined by denoting $\mu \leq_{\mathrm{cx}} \nu$ (resp. $\mu \leq_{\mathrm{cv}} \nu$, $\mu \leq_{\mathrm{icx}} \nu$ $\mu \leq_{\mathrm{icv}} \nu$) if (8) holds for all convex (resp. concave, increasing convex, increasing concave) functions ϕ. For random numbers X and Y distributed according to μ and ν, we denote $X \leq_{\mathrm{st}} Y$ if $\mu \leq_{\mathrm{st}} \nu$, and similarly for other integral stochastic orders.

When $X \leq_{\mathrm{st}} Y$ we say that X is smaller than Y in the strong order because then $\mathbb{P}(X > t) \leq \mathbb{P}(Y > t)$ for all t. When $X \leq_{\mathrm{cx}} Y$ we say that X is less variable than Y in the convex order, because then $\mathbb{E}X = \mathbb{E}Y$ and $\mathrm{Var}(X) \leq \mathrm{Var}(Y)$ whenever the second moments exist. Note that $X \leq_{\mathrm{cv}} Y$ if and only if $X \geq_{\mathrm{cx}} Y$, that is, X is less concentrated than Y. The order $X \leq_{\mathrm{icv}} Y$ can be interpreted by saying that X is smaller and less concentrated than Y.

3.2   Stochastic Ordering and Branching Processes

To obtain sharp results for branching processes, it is useful to introduce one more integral stochastic order. For probability distributions μ and ν on $\mathbb{R}_+$ (or a subset thereof), the *Laplace transform order* is defined by denoting $\mu \leq_{\mathrm{Lt}} \nu$ if (8) holds for all functions ϕ of the form $\phi(x) = -e^{-tx}$ with $t \geq 0$. Observe that $\mu \leq_{\mathrm{Lt}} \nu$ is equivalent to requiring $L_\mu(t) \geq L_\nu(t)$ for all $t \geq 0$, where we denote the Laplace transform of μ by $L_\mu(t) = \int e^{-tx}\mu(dx)$. For probability distributions p and q on $\mathbb{Z}_+$, observe that $p \leq_{\mathrm{Lt}} q$ if and only if their generating functions are ordered by $G_p(s) \geq G_q(s)$ for all $s \in [0,1]$. Because for any $t \geq 0$, the function $x \mapsto -e^{-tx}$ is increasing and concave, it follows that

$$\mu \leq_{\mathrm{st}} \nu \implies \mu \leq_{\mathrm{icv}} \nu \implies \mu \leq_{\mathrm{Lt}} \nu.$$

Due to the above implications we may interpret $X \leq_{\mathrm{Lt}} Y$ as X being smaller and less concentrated than Y (in a weaker sense than $X \leq_{\mathrm{icv}} Y$).

The following elementary result confirms an intuitive fact that a branching population with a smaller and more variable offspring distribution is less likely to survive in the long run. The proof can be obtained as a special case of a slightly more general result below (Lemma 2).

Proposition 2. *When* $p \leq_{Lt} q$, *the survival probabilities defined by* (5) *are ordered according to* $\zeta(p) \leq \zeta(q)$. *Especially,*

$$p \leq_{st} q \text{ or } p \leq_{cv} q \implies p \leq_{icv} q \implies p \leq_{Lt} q \implies \zeta(p) \leq \zeta(q).$$

4 Stochastic Ordering of the Configuration Model

Basic intuition about standard branching processes, as confirmed by Proposition 2, suggests that a large configuration model with a smaller and more variable degree distribution should have a smaller giant component. The next subsection displays a counterexample where this intuitive reasoning fails.

4.1 A Counterexample

Consider degree distributions p and q defined by

$$p = \frac{1}{8}\delta_1 + \frac{6}{8}\delta_2 + \frac{1}{8}\delta_3,$$
$$q = \frac{1}{16}\delta_0 + \frac{1}{8}\delta_1 + \frac{5}{8}\delta_2 + \frac{1}{8}\delta_3 + \frac{1}{16}\delta_4,$$

where δ_k represents the Dirac point mass at point k. Their downshifted size biasings, computed using (4), are given by

$$p^\circ = \frac{1}{16}\delta_0 + \frac{12}{16}\delta_1 + \frac{3}{16}\delta_2,$$
$$q^\circ = \frac{1}{16}\delta_0 + \frac{10}{16}\delta_1 + \frac{3}{16}\delta_2 + \frac{2}{16}\delta_3.$$

By comparing integrals of cumulative distributions functions [11, Theorem 3.A.1] or by constructing a martingale coupling [6], it is not hard to verify that in this case $p \leq_{cx} q$. Numerically computed values for the associated means, variances, and extinction probabilities are listed in Table 1. By evaluating the associated generating functions at $\eta(p^\circ) = 0.333$ and $\eta(q^\circ) = 0.186$, we find using (6) that $\zeta_{CM}(p) = 0.870$ and $\zeta_{CM}(q) = 0.892$.

This example shows that a standard branching process with a less variable offspring distribution ($p \leq_{cx} q$) is less likely to go extinct ($\eta(p) < \eta(q)$), but the same is not true for the downshifted size-biased offspring distributions

Table 1. Statistical indices associated to p and q and their downshifted size biasings.

	p	q	p°	q°
Mean	2.000	2.000	1.125	1.375
Variance	0.250	0.750	0.234	0.609
Extinction probability η	0.000	0.076	0.333	0.186

$(\eta(p^\circ) > \eta(q^\circ))$. As a consequence, the giant component of a large random graph corresponding to a configuration model with limiting degree distribution p is with high probability smaller than the giant component in a similar model with limiting degree distribution q, even though p is less variable than q. The reason for this is that, even though higher variability has an unfavorable effect on standard branching (the immediate neighborhood of the root note), higher variability also causes the neighbors of a neighbor to have bigger degrees on average.

4.2   A Monotonicity Result When One Extinction Probability is Small

The following result shows that increasing the variability of a degree distribution p *does* decrease the limiting relative size of a giant component, under the extra conditions that $p(0) = q(0)$ and that the extinction probability related to q° is less than $e^{-2} \approx 0.135$. Note that in the analysis of configuration models it is often natural to assume that $p(0) = q(0)$ because nodes without any half-links have no effect on large components.

Theorem 1. *Assume that $p \leq_{\text{icv}} q$, $p(0) = q(0)$, and $\eta(q^\circ) \leq e^{-2}$. Then $\zeta_{\text{CM}}(p) \leq \zeta_{\text{CM}}(q)$.*

Remark 1. Assume that $q(1) > 0$ and that $\zeta_{\text{CM}}(q) \geq 1 - q(0) - q(1)e^{-2}$. If this holds, then the inequality $G_q(s) \geq q(0) + q(1)s$ applied to $s = \eta(q^\circ)$ implies that

$$q(0) + q(1)e^{-2} \geq 1 - \zeta_{\text{CM}}(q) = G_q(\eta(q^\circ)) \geq q(0) + q(1)\eta(q^\circ),$$

so that $\eta(q^\circ) \leq e^{-2}$.

The proof of Theorem 1 is based on the following two lemmas.

Lemma 1. *If $p \leq_{\text{icv}} q$ and $p(0) = q(0)$, then the generating functions of the downshifted size biasings of p and q are ordered by*

$$G_{p^\circ}(s) \geq G_{q^\circ}(s) \quad \text{for all } s \in [0, e^{-2}].$$

Proof. Fix $s \in (0, e^{-2}]$, define a function $\phi : \mathbb{R}_+ \to \mathbb{R}_+$ by

$$\phi(x) = xs^x,$$

and observe that

$$G_{p^*}(s) = \frac{\mathbb{E}Xs^X}{\mathbb{E}X} = \frac{\mathbb{E}\phi(X)}{\mathbb{E}X}, \tag{9}$$

where X is a random integer distributed according to p. Denote $t = -\log s$, so that $t \in [2, \infty)$. Because $\phi'(x) = (1 - tx)e^{-tx}$ and $\phi''(x) = (tx - 2)te^{-tx}$, we find that ϕ is decreasing on $[\frac{1}{t}, \infty)$ and convex on $[\frac{2}{t}, \infty)$. Because $t \geq 2$, it follows that ϕ is decreasing and convex on $[1, \infty)$.

Now fix a decreasing convex function $\psi : \mathbb{R}_+ \to \mathbb{R}_+$ such that $\psi(x) = \phi(x)$ for all $x \geq 1$. Such a function can be constructed by letting ψ be linear on $[0, 1]$ and choosing the intercept and slope so that $\psi(1) = \phi(1)$ and $\psi'(1) = \phi'(1)$ (see Fig. 1). Let X^* and Y^* be some random integers distributed according to p^* and q^*, respectively. Because $\phi(0) = 0$, we see with the help of (9) that

$$G_{p^*}(s) = \frac{\mathbb{E}\phi(X)\mathbb{1}(X \geq 1)}{\mathbb{E}X} = \frac{\mathbb{E}\psi(X)\mathbb{1}(X \geq 1)}{\mathbb{E}X} = \frac{-\psi(0)p(0) + \mathbb{E}\psi(X)}{\mathbb{E}X}.$$

Observe now that $p \leq_{\mathrm{icv}} q$ implies that $\mathbb{E}X \leq \mathbb{E}Y$ and $\mathbb{E}\psi(X) \geq \mathbb{E}\psi(Y)$. Because $p(0) = q(0)$, it follows that

$$G_{p^*}(s) = \frac{-\psi(0)p(0) + \mathbb{E}\psi(X)}{\mathbb{E}X} \geq \frac{-\psi(0)q(0) + \mathbb{E}\psi(Y)}{\mathbb{E}Y} = G_{q^*}(s).$$

Because $G_{p^\circ}(s) = s^{-1}G_{p^*}(s)$ for $s \in (0, 1)$, we find that $G_{p^\circ}(s) \geq G_{q^\circ}(s)$ for all $s \in (0, e^{-2}]$. The claim is true also for $s = 0$, by the continuity of G_{p° and G_{q°.

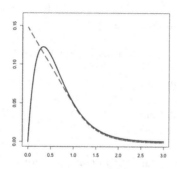

Fig. 1. Function ϕ (blue) and the its convex modification ψ (red) for $t = 3$ (Color figure online).

Lemma 2. *If $G_p(s) \geq G_q(s)$ for all $s \in [0, \eta(q)]$, then $\eta(p) \geq \eta(q)$.*

Proof. The claim is trivial for $\eta(q) = 0$, so let us assume that $\eta(q) > 0$. Then $G_q(0) > 0$, and the continuity of $s \mapsto G_q(s) - s$ implies that $G_q(s) > s$ for all $s \in [0, \eta(q))$. Hence also

$$G_p(s) \geq G_q(s) > s$$

for all $s \in [0, \eta(q))$. This shows that G_p has no fixed points in $[0, \eta(q))$ and therefore $\eta(p)$, the smallest fixed point of G_p in $[0, 1]$, must be greater than or equal to $\eta(q)$.

Proof (of Theorem 1). By applying Lemma 1 we see that

$$G_{p^\circ}(s) \geq G_{q^\circ}(s) \tag{10}$$

for all $s \in [0, e^{-2}]$. The assumption $\eta(q^\circ) \leq e^{-2}$ further guarantees that (10) is true for all $s \in [0, \eta(q^\circ)]$. Lemma 2 then shows that $\eta(p^\circ) \geq \eta(q^\circ)$. Finally,

$p \leq_{\mathrm{icv}} q$ implies $p \leq_{\mathrm{Lt}} q$, so that $G_p(s) \geq G_q(s)$ for all $s \in [0,1]$. Therefore, the monotonicity of G_p implies that

$$G_p(\eta(p^\circ)) \geq G_p(\eta(q^\circ)) \geq G_q(\eta(q^\circ)).$$

By substituting the above inequality into (6), we obtain Theorem 1.

4.3   Application to Social Network Modeling

Consider a large online social network of mean degree λ_0 where users forward copies of messages to their neighbors independently of each other with probability r_0. Without any a priori information about the higher order statistics of the degree distribution, one might choose to model the network using a configuration model with some degree distribution which is similar to one observed in some known social network. Because several well-studied social networks data exhibit a power-law tail in their degree data, a natural first choice is to model the unknown network using a configuration model with a Pareto-mixed Poisson limiting degree distribution (see Example 3)

$$p_0 = \mathrm{MPoi}(\mathrm{Par}(\alpha, \lambda_0(1 - 1/\alpha))) \tag{11}$$

with shape $\alpha > 1$ and mean λ_0.

Because the above choice of degree distribution was made without regard to network data, it is important to try to analyze how big impact can a wrong choice make to key network characteristics. When interested in global effects on information spreading, it is natural to consider the epidemic generated graph obtained by deleting stubs of the initial configuration model independently with probability $1 - r_0$. The outcome corresponds to another configuration model where the limiting degree p is the r_0-thinning of p_0, that is, the distribution of the random integer $X = \sum_{i=1}^{X_0} \theta_i$ with $X_0, \theta_1, \theta_2, \ldots$ being independent random integers such that X_0 is distributed according to p_0, and θ_i has the Bernoulli distribution with success probability r_0. Using generating functions one may verify that the r-thinning of a μ-mixed Poisson distribution $\mathrm{MPoi}(\mu)$ equals $\mathrm{MPoi}(r\mu)$, where $r\mu$ denotes the distribution of a μ-distributed random number multiplied by $r \in [0,1]$. Because $r\,\mathrm{Par}(\alpha, c) = \mathrm{Par}(\alpha, rc)$, it follows that the r_0-thinning of p_0 in (11) equals

$$p = \mathrm{MPoi}(\mathrm{Par}(\alpha, \lambda(1 - 1/\alpha))) \tag{12}$$

with $\lambda = \lambda_0 r_0$.

Now the key quantity describing the information spreading dynamics of the social network model is given by $\zeta_{\mathrm{CM}}(p)$ defined in (6). To study how sensitive this functional is to the variability of p, we have numerically evaluated $\zeta_{\mathrm{CM}}(p)$ for different values of α and λ, see Fig. 2. An extreme case is obtained by letting $\alpha \to \infty$ which leads to the standard Poisson distribution with mean λ. The dots on the right of Fig. 2 display the values of $\zeta_{\mathrm{CM}}(\mathrm{Poi}(\lambda))$. Again, perhaps a bit surprisingly, we see that for small values of λ, a Pareto-mixed Poisson as

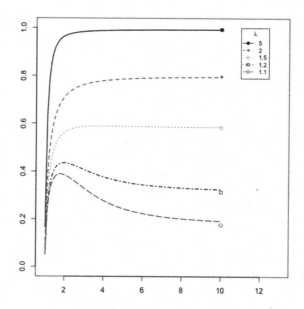

Fig. 2. Configuration model branching functional $\zeta_{\mathrm{CM}}(p_\alpha)$ for Pareto-mixed Poisson degree distribution with mean λ as a function of the tail exponent $\alpha > 1$.

a limiting degree distribution may produce an asymptotically larger maximally connected component in a configuration model than a one with a less variable unmixed Poisson distribution with the same mean. On the other hand, for larger values of λ, this phenomenon appears not to take place.

Proving the monotonicity of $\zeta_{\mathrm{CM}}(p)$ for Pareto-mixed Poisson distributions of the form (12) is not directly possible using Theorem 1 because $p(0)$ is not constant with respect to the shape parameter α. However, the following result can be applied here. Let us define a constant

$$\lambda_{\mathrm{cr}} = \inf\{\lambda \geq 0 : \lambda\zeta(\mathrm{Poi}(\lambda)) = 2\}.$$

Because $\lambda \mapsto \lambda\zeta(\mathrm{Poi}(\lambda))$ is increasing and continuous and grows from zero to infinity as λ ranges from zero to infinity, it follows that $\lambda_{\mathrm{cr}} \in (2, \infty)$ is well defined. Numerical computations indicate that $\lambda_{\mathrm{cr}} \approx 2.3$. The following result establishes a monotonicity result for the configuration model with a Pareto-mixed Poisson limiting distribution $p_\alpha = \mathrm{MPoi}(\mu_\alpha)$ with $\mu_\alpha = \mathrm{Par}(\alpha, c_\alpha)$, where the scale $c_\alpha = \lambda(1 - 1/\alpha)$ is chosen so that the mean of p_α equals λ for all $\alpha > 0$ (see Example 2).

Theorem 2. *For any $\lambda > \lambda_{\mathrm{cr}}$ there exists a constant $\alpha_{\mathrm{cr}} > 1$ such that*

$$\zeta_{\mathrm{CM}}(p_\alpha) \leq \zeta_{\mathrm{CM}}(p_\beta) \leq \zeta_{\mathrm{CM}}(\mathrm{Poi}(\lambda))$$

for all $\alpha_{\mathrm{cr}} \leq \alpha \leq \beta$.

Remark 2. Note that $\zeta_{CM}(\mathrm{Poi}(\lambda)) = \zeta(\mathrm{Poi}(\lambda))$ due to the fact that the Poisson distribution is invariant to downshifted size biasing (cf. Example 3).

Proof. Fix $\lambda > \lambda_{cr}$ and denote $\eta_\infty = \eta(\mathrm{Poi}(\lambda))$. Because $\lambda > \lambda_{cr}$, it follows that $\lambda(1 - \eta_\infty) > 2$, and therefore

$$\lambda(1 - \eta_\infty) \geq \frac{2}{1 - 1/\alpha_0} + \lambda\epsilon \tag{13}$$

for some large enough $\alpha_0 > 1$ and small enough $\epsilon > 0$. Next, Lemma 4 below shows that $\mu_\alpha^* = \mathrm{Par}(\alpha - 1, c_\alpha) \to \delta_\lambda$ and hence also $p_\alpha^\circ = \mathrm{MPoi}(\mu_\alpha^*) \to \mathrm{Poi}(\lambda)$ in distribution as $\alpha \to \infty$. The continuity of the standard branching functional implies that $\eta(p_\alpha^\circ) \to \eta_\infty$, and we may choose a constant $\alpha_{cr} \geq \alpha_0$ such that $\eta(p_\alpha^\circ) \leq \eta_\infty + \epsilon$ for all $\alpha \geq \alpha_{cr}$.

Assume now that $\alpha_{cr} \leq \alpha \leq \beta$. Then by [11, Theorem 3.A.5], one can check that

$$\mu_\alpha \leq_{cv} \mu_\beta \leq_{cv} \delta_\lambda. \tag{14}$$

Furthermore, $c_{\alpha_0} \leq c_\alpha \leq c_\beta$ implies that the supports of μ_α, μ_β, and δ_λ are contained in $[c_{\alpha_0}, \infty)$. Lemma 3 below implies that $G_{p_\alpha^\circ}(s) \geq G_{p_\beta^\circ}(s) \geq G_{\mathrm{Poi}(\lambda)}$ for all $s \in [0, s_0]$ where $s_0 = 1 - 2/c_{\alpha_0}$. Now (13) shows that

$$s_0 = 1 - \lambda^{-1}\left(\frac{2}{1 - 1/\alpha_0}\right) \geq 1 - \lambda^{-1}\left(\lambda(1 - \eta_\infty) - \lambda\epsilon\right) = \eta_\infty + \epsilon,$$

and hence the interval $[0, s_0]$ contains both $[0, \eta_\infty]$ and $[0, \eta(p_\beta^\circ)]$. By applying Lemma 2 twice, it follows that $\eta(p_\alpha^\circ) \geq \eta(p_\beta^\circ) \geq \eta(\mathrm{Poi}(\lambda)) = \eta_\infty$.

On the other hand, inequality (14) together with [11, Theorem 8.A.14] implies that $\mathrm{MPoi}(\mu_\alpha) \leq_{icv} \mathrm{MPoi}(\mu_\beta) \leq_{icv} \mathrm{Poi}(\lambda)$. Especially, $p_\alpha \leq_{Lt} p_\beta \leq_{Lt} \mathrm{Poi}(\lambda)$, so that $G_{p_\alpha} \geq G_{p_\beta} \geq G_{\mathrm{Poi}(\lambda)}$ pointwise on $[0, 1]$. This together with the monotonicity of the generating functions shows that

$$G_{p_\alpha}(\eta(p_\alpha^\circ)) \geq G_{p_\beta}(\eta(p_\beta^\circ)) \geq G_{\mathrm{Poi}(\lambda)}(\eta(\mathrm{Poi}(\lambda))),$$

and the claim follows by substituting the above inequalities into (6).

Lemma 3. *Let* $p = \mathrm{MPoi}(\mu)$ *and* $q = \mathrm{MPoi}(\nu)$ *where* $\mu \leq_{icv} \nu$. *Assume that the supports of* μ *and* ν *are contained in an interval* $[c, \infty)$ *for some* $c \geq 2$. *Then* $G_{p^\circ}(s) \geq G_{q^\circ}(s)$ *for all* $s \in [0, 1 - 2/c]$.

Proof. Note first that for $G_{\mathrm{MPoi}(\mu)}(s) = L_\mu(1 - s)$ and recall from Example 3 that $\mathrm{MPoi}(\mu)^\circ = \mathrm{MPoi}(\mu^*)$. Hence $G_{p^\circ}(s) = L_{\mu^*}(1 - s)$. Fix $s \in [0, 1 - 2/c]$ and note that $G_{p^\circ}(s) = m_1(\mu)^{-1} \int \phi_s(x)\, \mu(dx)$, where $\phi_s(x) = xe^{-(1-s)x}$. Because $\phi_s'(x) = (1 - (1-s)x)e^{-(1-s)x}$ and $\phi_s''(x) = (1-s)((1-s)x - 2)e^{-(1-s)x}$, it follows that the function ϕ_s is decreasing on $[\frac{1}{1-s}, \infty)$ and convex on $[\frac{2}{1-s}, \infty)$. Because $s \in [0, 1 - 2/c]$, it follows that ϕ_s is decreasing and convex on the support of μ_i for both $i = 1, 2$. Therefore $\mu \leq_{icv} \nu$ implies $\int \phi_s d\mu \geq \int \phi_s d\nu$. Because $\mu \leq_{icv} \nu$ also implies that the first moments are ordered according to $m_1(\mu) \leq m_1(\nu)$, we conclude that

$$G_{p^\circ}(s) = m_1(\mu)^{-1} \int \phi_s\, d\mu \geq m_1(\nu)^{-1} \int \phi_s\, d\nu = G_{q^\circ}(s).$$

Lemma 4. *If $c_\alpha \to \lambda \geq 0$ as $\alpha \to \infty$, then $\mathrm{Par}(\alpha, c_\alpha) \to \delta_\lambda$.*

Proof. Let U be a uniformly distributed random number in $(0,1)$. Then $X_\alpha = c_\alpha (1 - U)^{-1/\alpha}$ has $\mathrm{Par}(\alpha, c_\alpha)$ distribution for all α. Because $c_\alpha \to \lambda$ and $(1 - U)^{-1/\alpha} \to 1$, it follows that $X_\alpha \to \lambda$ almost surely, and hence also in distribution.

5   Conclusions

In this paper we studied the effect of degree variability to the global connectivity properties of large network models. The analysis was restricted to the configuration model and the associated uniform random graph with a given limiting degree distribution. Counterexamples were discovered both for a bounded support and power-law case that described that due to size biasing effects, increased degree variability may sometimes have a favorable effect on the size of the giant component, in sharp contrast to standard branching processes. We also proved using rigorous mathematical arguments that for some instances of strongly supercritical networks the increased degree variability has a negative effect on the global connectivity. To investigate whether or not this finding can be generalized outside the class of Pareto-mixed Poisson distributions is an interesting open problem for future research.

References

1. Ball, F.G., Sirl, D.J., Trapman, P.: Epidemics on random intersection graphs. Ann. Appl. Probab. **24**(3), 1081–1128 (2014). http://dx.doi.org/10.1214/13-AAP942
2. Bollobás, B., Janson, S., Riordan, O.: Sparse random graphs with clustering. Random Struct. Algorithms **38**(3), 269–323 (2011). http://dx.doi.org/10.1002/rsa.20322
3. Grimmett, G.R., Stirzaker, D.R.: Probability and Random Processes, 3rd edn. Oxford University Press, Oxford (2001)
4. van der Hofstad, R.: Random Graphs and Complex Networks. Lecture Notes, vol. I (2014). http://www.win.tue.nl/~rhofstad/NotesRGCN.html
5. Janson, S., Luczak, M.J.: A new approach to the giant component problem. Random Struct. Algor. **34**(2), 197–216 (2009). http://dx.doi.org/10.1002/rsa.20231
6. Leskelä, L., Vihola, M.: Conditional convex orders and measurable martingale couplings (2014). arXiv:1404.0999
7. Molloy, M., Reed, B.: A critical point for random graphs with a given degree sequence. Random Struct. Algor. **6**(2–3), 161–180 (1995). http://dx.doi.org/10.1002/rsa.3240060204
8. Molloy, M., Reed, B.: The size of the giant component of a random graph with a given degree sequence. Comb. Probab. Comput. **7**(3), 295–305 (1998). http://dx.doi.org/10.1017/S0963548398003526
9. Müller, A., Stoyan, D.: Comparison Methods for Stochastic Models and Risks. Wiley, New York (2002)
10. Sawaya, S., Klaere, S.: Extinction in a branching process: why some of the fittest strategies cannot guarantee survival. J. Stat. Distrib. Appl. **1**(10) (2014)
11. Shaked, M., Shanthikumar, J.G.: Stochastic Orders. Springer, New York (2007)
12. Valdés, J.E., Yera, Y.G., Zuaznabar, L.: Bounds for the expected time to extinction and the probability of extinction in the Galton-Watson process. Commun. Stat. - Theory Methods **43**(8), 1698–1707 (2014). http://dx.doi.org/10.1080/03610926.2012.673851

Navigability is a Robust Property

Dimitris Achlioptas[1] and Paris Siminelakis[2]([✉])

[1] Department of Computer Science, University of California, Santa Cruz, USA
optas@cs.ucsc.edu
[2] Department of Electrical Engineering, Stanford University, Stanford, USA
psimin@stanford.edu

Abstract. The Small World phenomenon has inspired researchers across a number of fields. A breakthrough in its understanding was made by Kleinberg who introduced Rank Based Augmentation (RBA): add to each vertex independently an arc to a random destination, selected from a carefully crafted probability distribution. Kleinberg proved that RBA makes many networks *navigable*, i.e., it allows greedy routing to successfully deliver messages between any two vertices in a polylogarithmic number of steps. Our goal in this work is to prove that navigability is an inherent, robust property of many random networks, requiring no augmentation, coordination, or even independence assumptions. Our framework assigns a cost to each edge and considers the uniform measure over all graphs on n vertices that satisfy a total budget constraint. We show that when the cost function is sufficiently correlated with the underlying geometry of the vertices and for a wide range of budgets, the overwhelming majority of all feasible graphs with the given budget are navigable. We provide a new set of geometric conditions that generalize Kleinberg's set systems as well as a unified analysis of navigability.

1 Introduction

The Small World phenomenon, popularly known as Six Degrees of Separation [21], refers to the fact that there exist short chains of acquaintances between most pairs of people in the world. Milgram's famous 1967 experiment [18] showed that not only such chains exist, but they can also be found in a decentralized manner. Specifically, each participant in the experiment was handed a letter addressed to a certain person and was told of some general characteristics of the person, including their occupation and location. They were then asked to forward the letter to the individual they knew on a first-name basis who was most likely to know the recipient. Based on the premise that similar individuals have higher chance of knowing each other (homophily), the participants typically forwarded the message to their contact most similar to the target, a strategy that

---

A longer version of this work appears at http://arxiv.org/abs/1501.04931.

D. Achlioptas—Research supported by ERC Starting Grant StG-210743 and an Alfred P. Sloan Fellowship.

P. Siminelakis—Supported in part by an Onassis Foundation Scholarship.

© Springer International Publishing Switzerland 2015
D.F. Gleich et al. (Eds.): WAW 2015, LNCS 9479, pp. 78–91, 2015.
DOI: 10.1007/978-3-319-26784-5_7

yielded remarkably short paths for most letters that reached their target (many did not).

In groundbreaking work [13,14], Kleinberg formulated mathematically the property of finding short-paths in a decentralized manner as *navigability*. Since then, the concept of navigability has also found applications in the design of P2P networks [6,23], data-structures [4,20] and search algorithms [17,22]. Key to decentralization is shared knowledge in the form of geometry, i.e., shared knowledge of a (distance) function on pairs of vertices (not necessarily satisfying the triangle inequality).

Geometry. *A geometry* (V, d) *consists of a set of vertices* V *and a distance function* $d : V \times V \to \mathbb{R}_+$, *where* $d(x, y) \geq 0$, $d(x, y) = 0$ *iff* $x = y$, *and* $d(x, y) = d(y, x)$, *i.e., the function* d *is a semi-metric.*

Given a graph[1] $G(V, E)$ on a geometry (V, d), a *decentralized search algorithm* is any algorithm[2] that given a target vertex t and current vertex v selects the next edge $\{v, u\} \in E$ to cross by only considering the distance of each neighbor u of v to the target t, i.e., $d(u, t)$. The allowance of paths of polylogarithmic length in the definition of navigability, below, is motivated by the fact that in any graph on n vertices with constant degree the diameter is $\Omega(\log n)$, reflecting an allowance for polynomial loss due to the lack of global information.

Navigability. *A graph* $G(V, E)$ *on geometry* (V, d) *is* d-*navigable if there exists a decentralized search algorithm which given any two vertices* $s, t \in V$ *will find a path from* s *to* t *of length* $O(\text{poly}(\log n))$.

In his original work on navigability [13], Kleinberg showed that if G is the 2-dimensional grid then adding a single random edge independently to each $v \in V$ results in a navigable graph (with d being the L1 distance on the grid). The distribution for selecting the other endpoint u of each added edge is crucial. Indeed, if it can only depend on $d(v, u)$ and distinct vertices are augmented independently, Kleinberg showed that there is a *unique* suitable distribution, the one in which the probability is proportional to $d(v, u)^{-2}$ (and, more generally, $d(v, u)^{-r}$ for r-dimensional lattices). The underlying principle behind Kleinberg's augmentation scheme has by now become known as *Rank Based Augmentation* (RBA) [14,16].

Rank Based Augmentation. *Given a geometry* (V, d), *a vertex* $v \in V$, *and* $\ell \geq 0$, *let* $N_v(\ell)$ *be the number of vertices within distance* ℓ *from* u. *In RBA, the probability of augmenting* v *with an edge to any* $u \in V$ *is*

$$P(v, u) \propto \frac{1}{N_u(d(v, u))}. \tag{1}$$

The intuition behind RBA is that navigability is attained because the added edges provide connectivity *across all distance scales*. Concretely, observe that for

---

[1] We use this notation instead of $G = (V, E)$ to avoid confusion between graphs and geometry.

[2] Here we are concerned mostly about the geometric-probabilistic requirements of Navigability and less on the performance and trade-offs between different decentralized algorithms. Thus, we will only analyze greedy routing.

any partition of the range of the distance function d into intervals, the probability that the augmenting edge is to a vertex at distance in a given interval is the same for every interval. Therefore, by partitioning the range of d into $O(\log n)$ intervals we see that, under RBA, whatever the current vertex v is there is always $\Omega((\log n)^{-1})$ probability that its long-range edge is to a vertex at the same "distance-scale" as the target. Of course, that alone is not enough. In order to shrink the distance to the target by a constant factor, we also need the long-range edge to have reasonable probability to go "in the right direction", something which is effortlessly true in regular lattices for any finite dimension. In subsequent work [14], aiming to provide rigorous results for graphs beyond lattices, Kleinberg showed that the geometric conditions needed for RBA to achieve navigability are satisfied by the geometries induced by *set-systems* satisfying certain conditions when the distance between two vertices is defined as the size of the smallest set (homophily) containing both.

Another canonical setting [7,9,15] for achieving navigability by RBA is when the distance function d is the shortest-path metric of a connected graph $G_0(V, E_0)$ with large diameter $\Theta(\text{poly}(n))$, also known as *graph augmentation*. In that setting, if E_d is the random set of edges added through RBA, the question is whether the (random) graph $G(V, E_0 \cup E_d)$ is d-navigable. Works of Slivkins [20] and Fraigniaud et al. [11] have shown the existence of a threshold, below which navigability is attainable and above which (in the worst case) it is not attainable, in terms of the *doubling dimension* of the shortest path metric of G_0. Roughly speaking, the doubling dimension corresponds to the logarithm of the possible directions that one might need to search, and the threshold occurs when it crosses $\Theta(\log \log n)$. Thus, we see that even when d is a (shortest path) metric, significant additional constraints on d need to be imposed.

The remarkable success of RBA in conferring navigability rests crucially on its *perfect adaptation* to the underlying geometry. This adaptation, though, requires not only all vertices to behave identically and independently, but also a very specific, indeed unique, functional form for the probability distribution of edge formation. This exact fine tuning renders RBA unnatural, severely undermining its plausibility. Our goal in this paper is to demonstrate that navigability is in fact a robust property of networks that does not require coordination or independence assumptions. We present arguments that point to the inevitable emergence of Navigability and of Rank Based Augmentation under the right geometric and "economical" conditions.

1.1 Related Work

The Small World phenomenon and Navigability are by now well studied topics. The review by Kleinberg [12] provides an excellent introduction and covers almost all of the earlier results up to 2006. Here, we would like to highlight two major questions that were left open and the work that has been made towards their resolution the past years.

Searchability of Networks. Duchon et al. [7] raised the question whether any graph $G(V, E_0)$ could become navigable after being augmented randomly with long range edges and proved that this is indeed possible for any graph of bounded growth. In the same direction, other authors have been looking at other general sufficient conditions on the underlying graph that enable navigability through augmentation. Fraigniaud [8] showed that this is possible for bounded-treewidth graphs, and Abraham et al. [1] showed it, further, for minor-free graphs. The work of Slivkins proved that augmentation always works if the doubling dimension of the graph is $O(\log \log n)$ and Fraigniaud et al. [11] proved that this is actually best possible. Since [11] research in this topic has turned to proving upper bounds for the performance of decentralized routing algorithms for arbitrary graphs. In that direction, Peleg first proved an $O(\sqrt{n})$ upper bound which was consequently improved to $O(n^{1/3})$ (up to poly-logarithmic factors) by Fraigniaud [9]. The best upper bound to date is $O(2^{(\log n)^{1/2+o(1)}})$ due to the work of Fraigniaud and Giakkoupis [10] almost matching a lower bound of $\Omega(2^{\sqrt{\log n}})$ for "monotone" decentralized algorithm by Fraigniaud et al. [11].

Evolution of Navigbability. Kleinberg's work identified a specific graph augmentation mechanism (RBA) that renders networks navigable, thus reducing the question of the evolution of Navigability to the evolution of Rank Based Augmentation. Fraigniaud et al. [5] and Sandberg [19] provided mechanisms based on random walks that aimed at providing an explanation of the latter. Specifically, Fraigniaud et al. [5] suggest that each vertex selects its long-range contact by performing a random walk with a restart probability harmonic with the number of steps performed. They show that the stationary distribution of this random walk is approximately the one given by RBA. On the other hand, Sandberg [19] considers the following evolutionary process, where source-destination pairs are continuously sampled and messages are forwarded through greedy routing. Each time a message fails to reach the destination within T times steps an edge is rewired between the source and destination pair. The author proves that the stationary distribution of the network exists and presents experimental evidence that suggest that the link distribution is similar to RBA.

2   Our Contribution

As mentioned, at the foundation of navigability lies shared knowledge in the form of geometry. At the same time, geometry imposes *global* constraints on the set of feasible networks. Most obviously, in a physical network where edges (wire, roads) correspond to a resource (copper, concrete) there is typically an upper bound on how much can be invested to create the network. More generally, cost may represent a number of different notions (e.g., class membership) that distinguish between edges.

   We will formalize the above intuition by (i) allowing edges to have costs given by an arbitrary function c on the edges, and (ii) taking as input an upper bound on the *total* cost of feasible graphs, i.e., a budget B. For instance, the cost of each edge may express the propensity (low cost) or reluctance (high cost) of two individuals interacting. In that case, an upper bound on the total cost expresses that feasible social interaction graphs are selected to not cause too much discomfort to the participating individuals.

Geometry, either of physical or of "concept" space, is an extremely natural backdrop for network formation that brings along both notions of cost and budget. In general, we expect that cost will correlate with geometry and that the budget, for any given cost function, will be such that the average degree of the network will be small (a property of nearly all real networks). Within these highly generic considerations, given a geometry, a cost function, and a budget we would like to study the set of all graphs satisfying the budget constraint, i.e., the set of all *feasible* graphs, and answer the following question: is it the case that the *overwhelming majority* is navigable?

This viewpoint departs from previous work where the aim was to provide a network creation *mechanism* that would lead to navigable graphs. Our viewpoint is motivated by the fact that, in reality, navigability is almost never an explicit goal of the network formation process yet, at the same time, navigability appears to be prevalent in a wide variety of settings. Roughly speaking, we isolate three ingredients that suffice for navigability on a geometry (V, d):

- Some degree of *coherence* of the distance function d (similar to Kleinberg's set systems).
- A *substrate* of connections between nearby points in V, making it impossible to get stuck.
- Sufficient, and sufficiently uniform, edge density across all distance scales.

The first two ingredients are generalizations of existing work and, as we will see, fully compatible with RBA. The third ingredient is also motivated by the RBA viewpoint, but we will prove that it can be achieved in far more-light handed, and thus natural, manner than RBA. Moreover, in the course of doing so, we will give RBA a very natural *economic* interpretation, as the distribution on edges arising when the cost of each edge is the *cost of indexing* among neighbours at the same distance scale.

Notation. Throughout the paper the set of vertices V is considered to be fixed and large, i.e., $n := |V|$ is finite but large. Any asymptotic notation, e.g. $f(|V|) = O(g(|V|))$ should be interpreted as comparing two functions of $|V|$ (eq. n) and only means that there are some constants independent of $|V|$ such that the corresponding inequalities hold, e.g. $f(|V|) \leq Cg(|V|)$. Lastly, to make the presentation more readable we will often say that a property $\mathcal{A}$ holds *with high probability (w.h.p)* to indicate that $\mathbb{P}(\mathcal{A}) \geq 1 - o(1)$.

2.1   Geometric Requirements and a Unifying Framework for RBA

We start by introducing the geometric requirements for navigability through the notion of *coherence*, that comes with an associated scale factor $\gamma > 1$. Specifically, given a geometry (V, d) we will refer to the vertices whose distance from a given vertex $v \in V$ lie in the interval $(\gamma^{k-1}, \gamma^k]$ as the vertices in the k-th (distance) γ-scale from v and denote their number as $P_k(v)$. Additionally for any two vertices $v \neq t \in V$ we will use k_{vt} to denote the integer k such that $d(v, t) \in (\gamma^{k-1}, \gamma^k]$. For a fixed $\lambda < 1$ and any target vertex $t \neq v$, we will say that a vertex u is t-helpful to v if $d(v, u) \leq \gamma^{k_{vt}}$ (u is within the same γ-scale as t from v), and $d(u, t) < \lambda d(v, t)$ (reduces the distance to t by a constant). We denote the set of t-helpful nodes of v by $D_\lambda(v, t)$.

Definition 1. *Fix $\gamma > 1$ and let $K = \lceil \log_\gamma |V| \rceil$. A geometry (V, d) is γ-coherent if:*

(H1) Bounded Growth: $\exists A > 1, \alpha \in (0, 1)$ *such that*

$$P_k(v) \in \gamma^k[\alpha, A], \, for \, all \, v \in V \, and \, k \in [K].$$

(H2) Isotropy: $\exists \phi > 0, 1 > \lambda > 0$ *such that*

$$|D_\lambda(v, t)| \geq \phi \gamma^{k_{vt}}, \, for \, all \, v \neq t \in V.$$

The two conditions above endow the, otherwise arbitrary, semi-metric d with sufficient regularity and consistency to guide the search. Although our definition of coherence is far more general, in order to convey intuition about the two conditions, think for a moment of V as a set of points in Euclidean space. The first condition guarantees that there are no "holes", as the variance in the density of points is bounded in every distance scale. The second condition guarantees that around any vertex v the density of points does not change much depending on the direction (target vertex t) and distance scale. Besides those two conditions, we make *no further* assumptions on d and, in particular, we do *not* assume the triangle inequality.

Coherent geometries allow us to provide a unified treatment of navigability since they encompass finite-dimensional lattices, hierarchical models, any vertex transitive graph with bounded doubling dimension and more generally as we show Kleinberg's set systems.

Theorem 1. *Every set system satisfying the conditions of [14] is a γ-coherent geometry for some $\gamma > 1$.*

Our second requirement is to assume the existence of a substrate, that implies that greedy routing will not get trivially stuck, i.e., that we can always move towards the target even incrementally.

Substrate. *A set of edges E_0 forms a substrate for a geometry (V, d), if for every $(s, t) \in V \times V$ with $s \neq t$, there is at least one vertex v such that $\{s, v\} \in E_0$ and $d(v, t) \leq d(s, t) - 1$. If there are multiple such vertices, we distinguish one arbitrarily and call it the* local t-connection *of s. A path starting from s and ending to t using only local t-connections is called a local (s, t)-path.*

In the graph augmentation setting this was given by the fact that the initial set of edges formed a known connected graph, while in Kleinberg's work on set systems it was circumvented by making the vertex degrees $\Theta(\log^2 n)$, so that the probability of ever being stuck at a vertex is polynomially small. We chose to use the notion of a substrate to encompass the graph augmentation setting but also generalize it since the semi-metric d is only *locally consistent* with the substrate. We show that those two requirements are sufficient for RBA to create a navigable graph.

Theorem 2. *Let (V, d) be any γ-coherent geometry and let E_0 be any substrate for it. If E_d is the (random) set of edges obtained by applying RBA to (V, d), then the graph $G(V, E_0 \cup E_d)$ is d-navigable w.h.p.*

Theorem 2 subsumes and unifies a number of previous positive results on RBA-induced navigability. Due to space limitations, the proofs of Theorems 1 and 2 are deferred to the full version of the paper. Our main contribution, though, lies in showing that given a substrate and coherence, navigability can emerge without any coordination or independence, merely from the alignment of cost and geometry.

2.2 Navigability from Organic Growth

As mentioned earlier, the success of RBA stems from the fact that the edge-creation mechanism is *perfectly* adapted to the underlying geometry so as to induce navigability. In contrast, we will not specify any edge-creation mechanism, but rather consider the set of *all* graphs feasible with a given budget. Our requirement is merely that the cost function is *informed* by the geometry, in the following sense.

γ-**consistency.** *Given a γ-coherent geometry (V, d), a cost function $c : V \times V \to \mathbb{R}$ is γ-consistent if c takes the same value c_k for every edge $\{u, v\}$ such that $d(u, v) \in (\gamma^{k-1}, \gamma^k]$.*

In other words, γ-consistency means that the partition of edges according to cost is a coarsening of the partition of the edges by γ-scale. Note that beyond γ-consistence we do not impose *any* constraint on the values $\{c_k\}$, not even a rudimentary one such as being increasing in k. In fact, even the γ-consistency requirement can be weakened significantly, as long as some correlation between the two partitions remains, but it is technically much simpler to assume γ-consistency as it greatly simplifies the exposition. One can think of consistency as limited sensitivity with respect to distance. As an example, it means that making friends with the people next door might be more likely than making friends with other people on the same floor, and that making friends with people on the same floor is more likely than making friends with people in a different floor, but it does not really matter which floor.

Cost-Geometries. *We say that $\Gamma = \Gamma(V, d, c)$ is a coherent cost-geometry if there exists $\gamma > 1$ such that (V, d) is a γ-coherent geometry and c is γ-consistent cost function.*

We are now in a position to state the set of feasible graphs that we consider.

Random Graphs of Bounded Cost. *Given a coherent cost-geometry $\Gamma(V, d, c)$ and a real number $B \geq 0$, let $G_\Gamma(B) = \{E \subseteq V \times V : \frac{1}{n}\sum_{e \in E} c(e) \leq B\}$, i.e., $G_\Gamma(B)$ is the set of all graphs (edge sets) on V with total cost at most Bn. A uniformly random element of $G_\Gamma(B)$ will be denoted as $E_\Gamma = E_\Gamma(B)$.*

Obtaining bounds on the probability that a uniformly random element out of $G_\Gamma(B)$ is navigable, is an intuitive and technically enabling way to obtain bounds on the fraction of feasible graphs that are navigable. Our main result is the following general theorem.

Theorem 3. *For every coherent cost-geometry $\Gamma(V, d, c)$ with substrate E_0, there exist numbers $B^{\pm}$ such that if E_Γ is a uniformly random element of $G_\Gamma(B)$ then:*

– *For all $B \leq B^+$, w.h.p. $|E_\Gamma| = O(n \cdot \text{poly}(\log n))$.*   *(Sparsity)*
– *For all $B \geq B^-$, w.h.p. the graph $G(V, E_0 \cup E_\Gamma)$ is d-navigable. (Navigability)*

Note that Theorem 3 shows that navigability arises eventually, i.e., for all $B \geq B^-$, without *any* further assumptions on the cost function or geometry. The caveat, if we think of B as increasing from 0, is that by the time there are enough edges across all distance scales, i.e., $B \geq B^-$, the total number of edges may be much greater than linear. This is because for an arbitrary cost structure $\{c_k\}$, by the time the "slowest growing" distance scale has the required number of edges, the other scales may be replete with edges, possibly many more than the required $\Omega(n/\text{poly}\log n)$. This is reflected in the ordering between B^- and B^+ that determines whether the sparsity and navigability regimes are overlapping. In particular, we would like $B^- \leq B^+$ and, ideally, the ratio $R = B^+/B^- > 0$ to be large. Whether this is the case or not depends precisely on the degree of adaptation of the cost-structure to the geometry, as we discuss next.

2.3 Navigability as a Reflection of the Cost of Indexing

Recall that for every vertex v in a γ-coherent geometry and for every distance scale $k \in [K]$, the number of vertices whose distance from v is in the k-th γ-distance scale is $P_k(v) = \Theta(\gamma^k)$. Let $p_k := \frac{1}{2|V|}\sum_{v \in V} P_k(v)$ be the average number of vertices at distance scale k from a random vertex. A coherent-cost geometry is parametrized by the numbers $\{p_k\}$ and the values of the cost function $\{c_k\}$.

We will now exhibit a class of cost functions that (i) have an intuitive interpretation as the average *cost of indexing*, (ii) achieve a ratio $R = B^+/B^- > 0$ that *grows* with n, i.e., a very wide range of budgets for which we have both navigability and sparsity, and (iii) recover RBA as a special case corresponding to a particular budget choice. To motivate the cost of indexing consider a vertex v that needs to forward a message to a neighbor u at the k-th distance scale. To do so, v needs to distinguish u among all other $P_k(v)$ vertices in the k-th distance scale, i.e., v needs to be able to *index* into that scale. Storing the unique ID of u among the other members of its equivalence class (in the eyes of v) has a cost of $\Theta(\log_2 P_k(v)) = \Theta(\log p_k) = \Theta(k)$ bits. Motivated by this we consider cost functions where for some $\beta > 0$,

$$c_k^* = \frac{1}{\beta} \log p_k. \tag{2}$$

Theorem 4. *For any coherent cost-geometry $\Gamma(V, d, c^*)$, there exist $B^\pm$ such that:*

(a) $B^+/B^- = \omega(\text{poly} \log n)$.
(b) For all $B \in [B^-, B^+]$, w.h.p. $|E_\Gamma(B)| = O(n \text{ poly} \log n))$ and the graph $G(V, E_0 \cup E_\Gamma(B))$ is d-navigable.
(c) There exists $B_\beta \in [B^-, B^+]$ such that Rank Based Augmentation is approximately recovered.

This concludes the presentation of our results. In the next two sections we outline the arguments that allow us to the prove our theorems.

3   Navigability via Reducibility and Uniform Richness

In this section we present structural results about navigability on coherent geometries that allow us to reduce navigability to a "richness" property of the probability measure on the non-substrate edges. We first define a sufficient deterministic property for navigability.

Reducibility. *If $G(V, E)$ is a graph on a coherent geometry (V, d) with substrate $E_0 \subseteq E$, we will say that $(s, t) \in V \times V$ is p-reducible if there is $C > 0$ such that among the first $C(\log |V|)^p$ vertices of the local (s, t)-path there is at least one vertex u such that $(u, v) \in E$ and $d(v, t) \leq \lambda d(s, t)$. If every pair $(s, t) \in V \times V$ is p-reducible, we will say that G is p-reducible.*

Reducibility expresses that as we move along the local path we never have to wait too long in order to encounter an edge that reduces the remaining distance by a constant factor. The motivation for introducing reducibility is that it allows us to separate the construction of the random graph from the analysis of the algorithm. Reducibility implies navigability in a straightforward manner.

Proposition 1. *If G is p-reducible, then greedy routing on G takes $O(\log^{1+p} n)$ steps.*

Reducibility is easiest to establish for random graphs whose edges are included independently, for concreteness we provide the following definition.

Product Measure. *Given a set of vertices V with $|V| = n$, let $\mathcal{G}_n$ denote the set of all $2^{\binom{n}{2}}$ possible graphs (edge-sets) on n vertices. A product measure on $\mathcal{G}_n$ is specified succinctly by a symmetric matrix $\mathbf{Q} \in [0, 1]^{n \times n}$ of probabilities where $Q_{ii} = 0$ for $i \in [n]$. We denote by $G(n, \mathbf{Q})$ the distribution over $\mathcal{G}_n$ in which possible edge $\{i, j\}$ is included independently with probability $Q_{ij} = Q_{ji}$.*

We next introduce the probabilistic requirement that suffices for reducibility.

Uniform Richness. *Let (V, d) be a γ-coherent geometry with parameter $\alpha \in (0, 1)$ (see H1). For $\theta \geq 1$, a product measure $G(n, \mathbf{Q})$ is θ-uniformly rich for (V, d) if there is a constant $M > 0$ such that for every $k \geq k_\theta$, for every pair (i, j) with $d(i, j) \in (\gamma^{k-1}, \gamma^k]$ we have:*

$$Q_{ij} \geq \frac{1}{M \log^\theta n} \frac{1}{\gamma^k}$$

where $k_\theta := \frac{\theta \log \log n - \log \alpha}{\log \gamma}$.

The number k_θ simply denotes the distance scale that would take $O(\log^\theta n)$ "slow" steps to cross, and is used to impose density requirements only for non-trivial distance scales as opposed to all scales. As we show next, uniform richness is a sufficient condition for reducibility on coherent geometries.

Lemma 1. *If (V, d) is a γ-coherent geometry with substrate E_0 and E_q is sampled from a θ-uniformly rich product measure $G(n, \mathbf{Q})$, then $G(V, E_0 \cup E_q)$ is $(\theta + 1)$-reducible with probability at least $1 - n^{-5}$.*

Deriving navigability from uniform richness may strike the reader as odd, given that a central goal of our work is to show that independence assumptions are *not* needed for navigability. There is no cause for alarm: we will never *assume* uniform richness. Instead, we will prove that under certain conditions, the (random) set of edges of a typical element of the set of all graphs feasible within a certain budget *dominates* a θ-uniformly rich product measure. Our capacity to do so is enabled by a very recent general theorem we developed in [2] which asserts that if a family of graphs $S \subseteq \mathcal{G}_n$ is sufficiently symmetric, then the uniform measure on S can be well-approximated by a product measure on the $\binom{n}{2}$ edges. We discuss this next.

4   Analyzing the Set of All Feasible Graphs

A classic result of random graph theory is that to study monotone properties of graphs with n vertices and m edges it suffices to study $G(n, p)$ random graphs, i.e., graphs generated by including each edge independently of all other with probability $p = p(m) = m/\binom{n}{2}$. The reason for this is that the uniform measure on graphs with exactly m edges is *sandwiched* by the $G(n, p(m))$ product measure, in the following sense.

Sandwichability. *The uniform measure $U(S)$ on an arbitrary set of graphs $S \subseteq \mathcal{G}_n$ is (ϵ, δ)-sandwichable if there exists a $n \times n$ symmetric matrix $\mathbf{Q}$ such that the two distributions $G^\pm \sim G(n, (1 \pm \epsilon)\mathbf{Q})$, and the distribution $G \sim U(S)$ can be coupled so that $G^- \subseteq G \subseteq G^+$ with probability at least $1 - \delta$.*

When S is the set of all graphs with exactly m edges we have $\mathbf{Q}_{ij} = p(m)$ for all non-diagonal entries. To make a sandwich, i.e., simultaneously generate G^-, G, G^+, one generates $\binom{n}{2}$ i.i.d. uniformly distributed real numbers in $[0, 1]$, one for each potential edge. The graph G^- contains all edges whose r.v. is less than $(1 - \epsilon)p$, the graph G contains the edges corresponding to the m smallest r.v.'s, while G^+ contains all edges whose r.v. is less than $(1+\epsilon)p$. As long as the m-th smallest r.v. is in $(1 - \epsilon, 1 + \epsilon)p$ we have $G^- \subseteq G \subseteq G^+$.

The set of all graphs with m edges is highly symmetric: its characteristic function is invariant under every permutation of the input $x \in \{0, 1\}^{\binom{n}{2}}$; it only cares about $|x|$. When considering graphs with bounded total cost, symmetry comes from the fact that edges with the same cost are interchangeable. Thus, if the number of distinct cost-classes is not too big we can hope for a product measure approximation (indeed, the set of all graphs with m edges can be seen as the case where there is only one cost class, unit cost, and the total budget is m). As discussed earlier, navigability requires some degree of structure in the underlying geometry in the form of coherence. Our requirement that the cost function is consistent with the (coherent) geometry, giving rise to a coherent cost-geometry, is what will give us enough symmetry to apply the main theorem of [2] and derive the following approximation.

In all of the following, $\Gamma(V, d, c)$ is an arbitrary coherent cost-geometry and $K = \lceil \log_\gamma |V| \rceil$. As before, we denote by c_k the cost of an edge of scale k and by p_k the average number of neighbors at distance scale k from a random vertex in V. For a given budget $B \geq 0$, let $\lambda(B) = g^{-1}(B) \geq 0$, where

$$g(\lambda) := \sum_{k=1}^{K} c_k \frac{p_k}{1 + \exp(\lambda c_k)}.$$

Intuitively, $\lambda(B)$ will control the drop in likelihood of costlier edges as a function of the budget B (mathematically, $\lambda(B)$ is a Lagrange multiplier, physically, it is an inverse temperature). The invertibility of g is as well as the proof of the following theorem will appear in the full version of the paper.

Theorem 5. *For every coherent cost-geometry Γ, there exists a constant $B_0(\Gamma) > 0$ such that for every $B \geq B_0(\Gamma)$ the uniform measure on $G_\Gamma(B)$ is (δ, ϵ)-sandwichable by the product measure $G(n, \mathbf{Q}^*(B))$ in which each edge of cost c_k has probability*

$$\mathbf{Q}^*_{ij}(B) = \frac{1}{1 + \exp(\lambda_\Gamma(B)c_k)}, \tag{3}$$

where $(\delta, \epsilon) = \left(\sqrt{\frac{24}{\log n}}, 2n^{-5K} \right)$.

Armed with Theorem 5 we can readily show that for a range of values of B, the product measure defined through (3) is θ-*uniformly rich* for some $\theta > 0$.

Proposition 2. *Let $\lambda_\theta(\{p_k\}, \{c_k\}) := \min_{k_\theta \leq k \leq K} \left[\frac{\log p_k}{c_k} \left(1 + \frac{\theta \log \log n}{\log p_k} \right) \right]$. Let $B_\theta^- := \max\{B_0(\Gamma), g(\lambda_\theta)\}$. For all $B \geq B_\theta^-$, the product measure $G(n, \mathbf{Q}^*(B))$ is θ-uniformly rich.*

Proof. This follows easily by the definition of λ_θ and the monotonicity of $\lambda(B) = g^{-1}(B)$ with respect to B. In particular, for any pair (i,j) of distance scale $k \geq k_\theta$ we have

$$Q_{ij}^*(B) = [1 + \exp(c_k \lambda(B))]^{-1} \geq \left[p_k \log^\theta(n)\right]^{-1} \geq \frac{1}{A \log^\theta(n) \gamma^k},$$

where the last inequality follows from (H1). $\qquad\square$

Proposition 3. *Let* $\Lambda_\theta(\{p_k\}, \{c_k\}) := \max_{k_\theta \leq k \leq K} \left[\frac{\log p_k}{c_k}\left(1 - \frac{\theta \log \log n}{\log p_k}\right)\right]$ *and* $B_\theta^+ := g(\Lambda_\theta)$. *For all* $B \leq B_\theta^+$, *the product measure* $G(n, \mathbf{Q}^*(B))$ *has* $O(n \log^{\theta+1} n)$ *edges with probability at least* $1 - n^{-5}$.

Proof. For all $B \leq B_\theta^+$, by definition of Λ_θ we have that for all $k \geq k_\theta$:

$$Q_{ij}^*(B) = [1 + \exp(c_k \lambda(B))]^{-1} \leq \left[p_k \log^{-\theta}(n)\right]^{-1}.$$

Thus, the *expected* number of edges $n \cdot \sum_{k=1}^K p_k [1 + \exp(\lambda(B)c_k)]^{-1}$ is upper bounded by

$$n \cdot \left[A k_\theta p_{k_\theta} + (K - k_\theta) \max_{k \geq k_\theta} p_k \frac{\log^\theta n}{p_k}\right] = n \cdot O\left(\log \log(n) \log^\theta n + \log(n) \log^\theta n\right),$$

since $k_\theta = O(\log \log n)$, $p_{k_\theta} = O(\log^\theta n)$ by (H1), and $K = O(\log n)$. Expressing the number of edges as a sum of independent Bernoulli random variables and applying standard Chernoff bounds [3] we get the required conclusion. $\qquad\square$

Proof of Theorem 3. For any $B \geq B_0(\Gamma)$, consider two random elements generated according to $E^\pm \sim G(n, (1 \pm \epsilon)\mathbf{Q}^*(B))$ and let W be the event that $E^- \subseteq E_\Gamma \subseteq E^+$. Theorem 5 implies that for $\epsilon = \sqrt{24/\log(n)}$ the probability of W is at least $1 - n^{-5K}$. Further, for any constant $p > 0$ and for an arbitrary set of edges E let $N_p(E)$ denote the event that the graph $G(V, E_0 \cup E))$ is not p-reducible and let $N_d(E)$ be the event that the same graph is not d-navigable. Since p-reducibility is a monotone increasing property with respect to edge inclusion and since $N_d \subseteq N_p$ by Proposition 1, we get

$$\mathbb{P}(N_d(E_\Gamma)) = \mathbb{P}(N_d(E_\Gamma) \cap W) + \mathbb{P}(N_d(E_\Gamma) \cap \overline{W}) \tag{4}$$
$$\leq \mathbb{P}(N_p(E_\Gamma)|W) + \mathbb{P}(\overline{W}) \tag{5}$$
$$\leq \mathbb{P}_{\mathbf{Q}^*}(N_p(E^-)) + 2n^{-5K} \tag{6}$$
$$\leq n^{-5} + 2n^{-5K}, \tag{7}$$

where we used the law of total probability in the first equality, Bayes Theorem in the second inequality, Theorem 5 and monotonicity of reducibility in the third. The last inequality follows from Lemma 1 and Proposition 2. This proves part (a) of the theorem. To prove part (b) we follow the same method but for the event $\{|E_\Gamma| = \omega(npolylog n)\}$ and exploit that, conditional on W occurring, $E_\Gamma \subseteq E^+$. Using Proposition 3 and Theorem 5 we get the required conclusion. $\qquad\square$

Proof Sketch of Theorem 4. Due to space constraints we only briefly mention that the proof of Theorem 4 follows by using the explicit form of the cost function in (2), the expressions for λ_θ and Λ_θ in Propositions 2 and 3 to prove parts (a) and (b), and finally the expression for Q_{ij}^* in (3) and property (H1) to prove part (c). □

References

1. Abraham, I., Gavoille, C., Malkhi, D.: Compact routing for graphs excluding a fixed minor. In: Fraigniaud, P. (ed.) DISC 2005. LNCS, vol. 3724, pp. 442–456. Springer, Heidelberg (2005)
2. Achlioptas, D., Siminelakis, P.: Symmetric graph properties have independent edges. In: Halldórsson, M.M., Iwama, K., Kobayashi, N., Speckmann, B. (eds.) ICALP 2015. LNCS, vol. 9135, pp. 467–478. Springer, Heidelberg (2015)
3. Alon, N., Spencer, J.: The Probabilistic Method. John Wiley, New York (1992)
4. Aspnes, J., Diamadi, Z., Shah, G.: Fault-tolerant routing in peer-to-peer systems. In: Proceedings of the 21st Annual ACM Symposium on Principles of Distributed Computing, PODC 2002, pp. 223–232 (2002)
5. Chaintreau, A., Fraigniaud, P., Lebhar, E.: Networks become navigable as nodes move and forget. In: Aceto, L., Damgård, I., Goldberg, L.A., Halldórsson, M.M., Ingólfsdóttir, A., Walukiewicz, I. (eds.) ICALP 2008, Part I. LNCS, vol. 5125, pp. 133–144. Springer, Heidelberg (2008)
6. Clarke, I., Sandberg, O., Wiley, B., Hong, T.W.: Freenet: a distributed anonymous information storage and retrieval system. In: Federrath, H. (ed.) Anonymity 2000. LNCS, vol. 2009, pp. 46–66. Springer, Heidelberg (2001)
7. Duchon, P., Hanusse, N., Lebhar, E., Schabanel, N.: Could any graph be turned into a small-world? Theor. Comput. Sci. **355**(1), 96–103 (2006)
8. Fraigniaud, P.: Greedy routing in tree-decomposed graphs. In: Brodal, G.S., Leonardi, S. (eds.) ESA 2005. LNCS, vol. 3669, pp. 791–802. Springer, Heidelberg (2005)
9. Fraigniaud, P., Gavoille, C., Kosowski, A., Lebhar, E., Lotker, Z.: Universal augmentation schemes for network navigability. Theor. Comput. Sci. **410**(21–23), 1970–1981 (2009)
10. Fraigniaud, P., Giakkoupis, G.: On the searchability of small-world networks with arbitrary underlying structure. In: Proceedings of the 42nd ACM Symposium on Theory of Computing, STOC 2010, pp. 389–398 (2010)
11. Fraigniaud, P., Lebhar, E., Lotker, Z.: A lower bound for network navigability. SIAM J. Discrete Math. **24**(1), 72–81 (2010)
12. Kleinberg, J.: Complex networks and decentralized search algorithms. In: Proceedings of the International Congress of Mathematicians, ICM 2006, pp. 1019–1044 (2006)
13. Kleinberg, J.M.: The small-world phenomenon: an algorithmic perspective. In: Proceedings of the 32nd Annual ACM Symposium on Theory of Computing, STOC 2000, pp. 163–170 (2000)
14. Kleinberg, J.M.: Small-world phenomena and the dynamics of information. In: Proceedings of Advances in Neural Information Processing Systems 14, NIPS 2001, pp. 431–438 (2001)
15. Lebhar, E., Schabanel, N.: Graph augmentation via metric embedding. In: Baker, T.P., Bui, A., Tixeuil, S. (eds.) OPODIS 2008. LNCS, vol. 5401, pp. 217–225. Springer, Heidelberg (2008)

16. Liben-Nowell, D., Novak, J., Kumar, R., Raghavan, P., Tomkins, A.: Geographic routing in social networks. Proc. Natl. Acad. Sci. U.S.A. **102**(33), 11623–11628 (2005)
17. Manku, G.S., Naor, M., Wieder, U.: Know thy neighbor's neighbor: the power of lookahead in randomized P2P networks. In: Proceedings of the 36th Annual ACM Symposium on Theory of Computing, STOC 2004, pp. 54–63 (2004)
18. Milgram, S.: The small world problem. Psychol. Today **2**(1), 60–67 (1967)
19. Sandberg, O.: Neighbor selection and hitting probability in small-world graphs. Ann. Appl. Probab. **18**(5), 1771–1793 (2008)
20. Slivkins, A.: Distance estimation and object location via rings of neighbors. Distrib. Comput. **19**(4), 313–333 (2007)
21. Watts, D.J.: Six Degrees: The Science of a Connected Age. WW Norton, New York (2004)
22. Zeng, J., Hsu, W.-J., Wang, J.: Near optimal routing in a small-world network with augmented local awareness. In: Pan, Y., Chen, D., Guo, M., Cao, J., Dongarra, J. (eds.) ISPA 2005. LNCS, vol. 3758, pp. 503–513. Springer, Heidelberg (2005)
23. Zhang, H., Goel, A., Govindan, R.: Using the small-world model to improve freenet performance. Comput. Netw. **46**(4), 555–574 (2004)

Dynamic Processes on Large Graphs

Local Majority Dynamics on Preferential Attachment Graphs

Mohammed Amin Abdullah[1]([✉]), Michel Bode[2], and Nikolaos Fountoulakis[2]

[1] Mathematical and Algorithmic Sciences Lab,
Huawei Technologies Ltd., Boulogne-Billancourt, France
mohammed.abdullah@huawei.com
[2] School of Mathematics, University of Birmingham, Birmingham, UK
michel.bode@gmx.de, n.fountoulakis@bham.ac.uk

Abstract. Suppose in a graph G vertices can be either red or blue. Let k be odd. At each time step, each vertex v in G polls k random neighbours and takes the majority colour. If it doesn't have k neighbours, it simply polls all of them, or all less one if the degree of v is even. We study this protocol on the preferential attachment model of Albert and Barabási [3], which gives rise to a degree distribution that has roughly power-law $P(x) \sim \frac{1}{x^3}$, as well as generalisations which give exponents larger than 3. The setting is as follows: Initially each vertex of G is red independently with probability $\alpha < \frac{1}{2}$, and is otherwise blue. We show that if α is sufficiently biased away from $\frac{1}{2}$, then with high probability, consensus is reached on the initial global majority within $O(\log_d \log_d t)$ steps. Here t is the number of vertices and $d \geq 5$ is the minimum of k and m (or $m-1$ if m is even), m being the number of edges each new vertex adds in the preferential attachment generative process. Additionally, our analysis reduces the required bias of α for graphs of a given degree sequence studied in [1] (which includes, e.g., random regular graphs).

Keywords: Local majority dynamics · Preferential attachment · Power-law graphs · Voting · Consensus

1 Introduction

Let $G = (V, E)$ be a graph where each vertex maintains an opinion, which we speak of in terms of two colours - red and blue. We make no assumptions about the properties of the colours/opinions except that vertices can distinguish between them. We are interested in distributed protocols on G that can bring about consensus to a single opinion.

One of the simplest and most widely studied distributed consensus algorithms is the *voter model* (see, e.g., [4, Chap. 14]). In the discrete time setting, at each time step τ, each vertex chooses a single neighbour uniformly at random (**uar**)

M.A. Abdullah and N. Fountoulakis—Research supported by the EPSRC Grant No. EP/K019749/1.

and assumes its opinion. The number of different opinions in the system is clearly non-increasing, and consensus is reached almost surely in finite, non-bipartite, connected graphs. Using an elegant martingale argument, [13] determined the probability of consensus to a particular colour. In our context this would be the sum of the degrees of vertices which started with that colour, as a fraction of the sum of degrees over all vertices. Thus, on regular graphs, for example, if the initial proportion of reds is a constant α, the probability of a red consensus is α. This probability is increased on non-regular graphs if the minority is "privileged" by sitting on high degree vertices (as in say, for example, the small proportion of high degree vertices in a graph with power-law distribution). This motivates an alternative where the majority is certain, or highly likely, to win.

The *local majority* protocol in a synchronous discrete time setting does the following: At each time step, each vertex v polls all its neighbours and assumes the majority colour in the next time step. This can be motivated by both a prescriptive and a descriptive view. In the former, as a consensus protocol, it can be seen as a distributed co-ordination mechanism for networked systems. In the latter, it can be seen as a natural process occurring, for example in social networks where it may represent the spread of influence.

Let k be odd. Suppose at time step $\tau = 0$ each vertex of a graph $G = (V, E)$ is either red or blue. In this paper we study the following generalisation of the local majority protocol (also in a synchronous, discrete time setting):

Definition 1 (k-choice Local Majority Protocol $\mathcal{MP}^k$). *For each vertex $v \in V$, for each time step $\tau = 1, 2, \ldots$ do the following: choose a set of k neighbours of v uniformly at random. The colour of v at time step τ is the majority colour of this set at time step $\tau - 1$. If v does not have k neighbours, then choose a random set of largest possible odd cardinality.*

Clearly, we can retrieve the local majority protocol by setting k to be the maximum degree, for example.

In addition to which colour dominates, one is also interested in how long it takes to reach consensus. In the voter model, there is a duality between the voting process and multiple random walks on the graph. The time it takes for a single opinion to emerge is the same as the time it takes for n independent random walks - one starting at each vertex - to coalesce into a single walk, where two or more random walks coalesce if they are on the same vertex at the same time. Thus, consensus time can be determined by studying this multiple walk process. However, the analyses of local-majority-type protocols have not been readily amenable to the established techniques for the voter model, namely, martingales and coalescing random walks. Martingales have proved elusive and the random walks duality does not readily transfer, nor is there an obvious way of altering the walks appropriately. Thus, ad-hoc techniques and approaches have been developed.

We say a sequence of events $(\mathcal{E}_t)_t$ occurs *with high probability* (**whp**) if $\mathbf{Pr}(\mathcal{E}_t) \to 1$ as $t \to \infty$. In this paper, the underlying parameter t which goes to infinity will be the number of vertices in the sequence of graphs $\mathrm{PA}_t(m, \delta)$ we consider.

The main result in this paper will be to show that when each vertex of a preferential attachment graph $\mathrm{PA}_t(m, \delta)$ (introduced in the next section) is red independently with probability $\alpha < 1/2$, where α is sufficiently biased away from $1/2$, then the system will converge to the majority colour with high probability, and we give an upper bound for the number of steps this takes.

2 Preferential Attachment Graphs

The preferential attachment models have their origins in the work of Yule [17], where a growing model is proposed in the context of the evolution of species. A similar model was proposed by Simon [16] in the statistics of language. The principle of these models was used by Albert and Barabási [3] to describe a random graph model where vertices arrive one by one and each of them throws a number of half-edges to the existing graph. Each half-edge is connected to a vertex with probability that is proportional to the degree of the latter. This model was defined rigorously by Bollobás, Riordan, Spencer and Tusnády [6] (see also [5]). We will describe the most general form of the model which is essentially due to Dorogovtsev et al. [11] and Drinea et al. [12]. Our description and notation below follows that from the book of van der Hofstad [14].

The random graph $\mathrm{PA}_t(m, \delta) = (V, E)$ where $V = [t]$ is parameterised by two constants: $m \in \mathbb{N}$, and $\delta \in \mathbb{R}$, $\delta > -m$. It gives rise to a random graph sequence (i.e., a sequence in which each member is a random graph), denoted by $(\mathrm{PA}_t(m, \delta))_{t=1}^{\infty}$. The tth term of the sequence, $\mathrm{PA}_t(m, \delta)$ is a graph with t vertices and mt edges. Further, $\mathrm{PA}_t(m, \delta)$ is a subgraph of $\mathrm{PA}_{t+1}(m, \delta)$. We define $\mathrm{PA}_t(1, \delta)$ first, then use it to define the general model $\mathrm{PA}_t(m, \delta)$ (the Barabási-Albert model corresponds to the case $\delta = 0$).

The random graph $\mathrm{PA}_1(1, \delta)$ consists of a single vertex with one self-loop. We denote the vertices of $\mathrm{PA}_t(1, \delta)$ by $\{v_1^{(1)}, v_2^{(1)}, \ldots, v_t^{(1)}\}$. We denote the degree of vertex $v_i^{(1)}$ in $\mathrm{PA}_t(1, \delta)$ by $D_i(t)$. Then, conditionally on $\mathrm{PA}_t(1, \delta)$, the growth rule to obtain $\mathrm{PA}_{t+1}(1, \delta)$ is as follows: We add a single vertex $v_{t+1}^{(1)}$ having a single edge. The other end of the edge connects to $v_{t+1}^{(1)}$ itself with probability $\frac{1+\delta}{t(2+\delta)+(1+\delta)}$, and connects to a vertex $v_i^{(1)} \in \mathrm{PA}_t(1, \delta)$ with probability $\frac{D_i(t)+\delta}{t(2+\delta)+(1+\delta)}$ – we write $v_{t+1}^{(1)} \to v_i^{(1)}$. For any $t \in \mathbb{N}$, let $[t] = \{1, \ldots, t\}$. Thus,

$$\mathbf{Pr}\left(v_{t+1}^{(1)} \to v_i^{(1)} \mid \mathrm{PA}_t(1, \delta)\right) = \begin{cases} \frac{1+\delta}{t(2+\delta)+(1+\delta)} & \text{for } i = t+1, \\ \frac{D_i(t)+\delta}{t(2+\delta)+(1+\delta)} & \text{for } i \in [t] \end{cases}$$

The model $\mathrm{PA}_t(m, \delta)$, $m > 1$, with vertices $\{1, \ldots, t\}$ is derived from $\mathrm{PA}_{mt}(1, \delta/m)$ with vertices $\{v_1^{(1)}, v_2^{(1)}, \ldots, v_{mt}^{(1)}\}$ as follows: For each $i = 1, 2, \ldots, t$, we contract the vertices $\{v_{(i-1)+1}^{(1)}, v_{(i-1)+2}^{(1)}, \ldots, v_{(i-1)+m}^{(1)}\}$ into one super-vertex, and identify this super-vertex as i in $\mathrm{PA}_t(m, \delta)$. When a contraction takes place, all loops and multiple edges are retained. Edges shared between a set of contracted vertices become loops in the contracted super-vertex. Thus, $\mathrm{PA}_t(m, \delta)$ is a graph on $[t]$.

The above process gives a graph whose degree distribution follows a power law with exponent $3 + \delta/m$. This was suggested by the analyses in [11,12]. It was proved rigorously for integral δ by Buckley and Osthus [7]. For a full proof for real δ see [14]. In particular, when $-m < \delta < 0$, the exponent is between 2 and 3. Experimental evidence has shown that this is the case for several networks that emerge in applications (cf. [3]). Furthermore, when $m \geq 2$, then $PA_t(m, \delta)$ is **whp** connected, but when $m = 1$ this is not the case, giving rise to a logarithmic number of components (see [14]).

3   Results and Related Work

Our main result is the following.

Theorem 1. *Let $k \geq 5$ be odd and let $d = \min\{m, k\}$ if m is odd and $d = (m - 1) \wedge k$ if m is even. Let α^* be the smallest positive solution for x in the equation $\mathbf{Pr}\left(Bin(d - 1, x) \geq \frac{d-1}{2}\right) = x$. If $\delta \geq 0$ and each vertex in $PA_t(m, \delta)$ is red independently with probability $\alpha < \alpha^*$, then for any constant $\varepsilon > 0$, **whp** under $\mathcal{MP}^k$ every vertex in $PA_t(m, \delta)$ is blue at all time steps $\tau \geq \frac{1+\varepsilon}{\log_d\left(\frac{d-1}{2}\right)} \log_d \log_d t$.*

Note that $\delta = 0$ gives the model proposed in the seminal work of Albert and Barabási [3], giving power law exponent 3, and that $\delta > 0$ gives exponents larger than 3. We refer the reader to [14] for further details.

Note, for $d = 5, 7, 9, 11$ we have $\alpha^* = 0.232, 0.347, 0.396, 0.421$ to 3 significant figures (s.f.), respectively. Of course, $\alpha^* \to \frac{1}{2}$ as $d \to \infty$.

The most closely related work is [1]. Here, the same protocol was studied on random graphs of a given degree sequence (which includes random regular graphs) and Erdős–Rényi random graphs slightly above the connectivity threshold. Results similar to Theorem 1 were obtained, and in the case of the former model, an almost matching lower bound was shown. It should be noted that the thresholds for α obtained in this work apply equally to the models in [1], and improve the thresholds for α. To contrast, in that paper, the thresholds for $d = 5, 7, 9, 11$ were $0.092, 0.182, 0.234, 0.268$ to 3 s.f., respectively.

In [15], (full) local majority dynamics on d-regular λ-expanders on n vertices is studied. In our notation, they show that when $\alpha \leq 1/2 - \frac{2\lambda n}{d}$, there is convergence to the initial majority, so long as $\frac{\lambda}{d} \leq \frac{3}{16}$. Since $\lambda \geq (1 - o(1))\sqrt{d}$ for a d-regular graph, this condition implies $d \geq 29$. In contrast, our results apply for $d \geq 5$.

In [10] a variant of local majority is studied where a vertex contacts m others and if d of them have the same colour, the vertex subsequently assumes this colour. They demonstrate convergence time of $O(\log n)$ and error probability – the probability of converging on the initial minority – decaying exponentially with n. However, the analysis is done only for the complete graph; our analysis of sparse graphs is a crucial difference, because the techniques employed for complete graphs do not carry through to sparse graphs, nor are they easily adapted. The error probability we give is not as strong but still strong, nevertheless. Furthermore, the convergence time we give is much smaller.

In [8] the authors study the following protocol on random regular graphs and regular expanders: Each vertex picks two neighbours at random, and takes the majority of these with itself. They show convergence to the initial majority in $O(\log n)$ steps with high probability, subject to sufficiently large initial bias and high enough vertex degree. However, in their setting, the placement of colours can be made adversarially.

In summary, our contribution is demonstrating convergence and time of convergence to initial majority for a generalisation of local majority dynamics for preferential attachment graphs with power-law exponent 3 and above. As far as we know this is the only such result for power-law graphs (by preferential attachment or otherwise). Furthermore, we have improved the bias thresholds for graphs of a given degree sequence studied by the first author in [1], which, to the best of our knowledge, were already the best or only known results for small degree graphs in this class (which includes, e.g., random regular graphs).

4   Structural Results

Throughout this paper we let $\gamma = \gamma(m, \delta) = \frac{1}{2+\delta/m}$. Observe the condition $\delta > -m$ (which must be imposed), implies $0 < \gamma < 1$.

Furthermore, for two non-negative functions $f(t), g(t)$ on $\mathbb{N}$ we write $f(t) \lesssim g(t)$ to denote that $f(t) = O(g(t))$. The underlying asymptotic variable will always be t, the number of vertices in $\mathrm{PA}_t(m, \delta)$.

Let A be a large constant and let

$$\omega = A \log \log t.$$

Let

$$\kappa = (\log t)^{7\omega}$$

and define as the *inner core* the vertices $[\kappa]$, and refer to them as *heavy* vertices. We also refer to vertices outside the inner core as *light* vertices.

Let

$$\kappa_o = (\log t)^{999\omega}$$

and define as the *outer core* the vertices $[\kappa_o]$.

Call a path *short* if it has length at most ω. Call a cycle *short* if it has at most $2\omega + 1$ vertices. Here "cycle" includes a pair of vertices connected by parallel edges and a vertex with a self-loop.

Below, we repeatedly apply the following, which is proved in [2] (and for the case $k = 1$ was given in [14]):

Proposition 1. *Suppose $i_1, j_1, i_2, j_2, \ldots, i_k, j_k$ are vertices in $\mathrm{PA}_t(m, \delta)$ where $i_s < j_s$ for $s = 1, 2, \ldots, k$. Then*

$$\mathbf{Pr}(j_1 \to i_2 \cap j_2 \to i_2, \ldots, j_k \to i_k) \leq M^k \frac{1}{i_1^\gamma j_1^{1-\gamma}} \frac{1}{i_2^\gamma j_2^{1-\gamma}} \cdots \frac{1}{i_k^\gamma j_k^{1-\gamma}}$$

where $M = M(m, \delta)$ is a constant that depends only on m and δ.

Below, we also use the fact that $\frac{1}{i^\gamma j^{1-\gamma}} \leq \frac{1}{(ij)^{\frac{1}{2}}}$ when $\delta \geq 0$ and $i \leq j$. A similar counting approach was used in [9].

For a vertex v define $\mathcal{B}(v, r)$ to be the r ball of v in $\mathrm{PA}_t(m, \delta)$, the subgraph within distance r.

Lemma 1. *With high probability, every vertex $v > \kappa$ has the following property: $\mathcal{B}(v, \omega)$ contains at most one cycle consisting entirely of light vertices.*

Proof. Define a *cycle-path-cycle* (CPC) structure as a pair of cycles connected by a path. We consider CPC structures where the cycles and paths are short, that is, cycles have sizes $1 \leq r, s \leq 2\omega + 1$, and the path has length $0 \leq \ell \leq \omega$. Note $r = 1$ denotes a self-loop and $r = 2$ denotes a pair of parallel edges between two vertices.

We denote by $a_1, \ldots, a_r$ and $b_1, \ldots, b_s$ the vertices of the cycles, and $c_0, \ldots, c_\ell$ the vertices of the path. Without loss of generality, we may assume $a_1 = c_0$ and $b_1 = c_\ell$. Thus, the structure has $r + s + \ell$ edges and $r + s + \ell - 1$ vertices.

Applying Proposition 1, the expected number of such structures lying entirely in $[t] \setminus [\kappa]$ is bounded by

$$\sum_{r=1}^{2\omega+1} \sum_{\ell=0}^{\omega-1} \sum_{s=1}^{2\omega+1} \sum_{\kappa < a_1, \ldots, a_r} \sum_{\kappa < b_1, \ldots, b_s} \sum_{\kappa < c_1, \ldots, c_{\ell-1}} \frac{M^{r+s+\ell}}{(a_1 b_1)^{3/2}} \prod_{i=2}^{r} \frac{1}{a_i} \prod_{j=2}^{s} \frac{1}{b_j} \prod_{k=1}^{\ell-1} \frac{1}{c_k}$$

$$\lesssim \left(\int_\kappa^t x^{-3/2} \, \mathrm{d}x \right)^2 \sum_{r=1}^{2\omega+1} \sum_{\ell=0}^{\omega-1} \sum_{s=1}^{2\omega+1} (M \log t)^{r+s+\ell}$$

$$\lesssim \frac{(\log t)^{6\omega}}{\kappa} = o(1).$$

The rest of the proof is of a similar nature and is given in the full version of the paper. $\qquad\square$

Lemma 2. *With high probability, the following hold for all $v > \kappa_0$:*

(i) *v has at most 2 edges on short paths into $[\kappa]$.*
(ii) *If v is on a short light cycle, then v has no edge that is on a short (light) path into $[\kappa]$ but that is not part of the cycle.*
(iii) *If v is connected to a short light cycle C by a short light path P, then v has at most one edge e such that e is on a short path into $[\kappa]$ but $e \notin P$.*

Proof. (i) Suppose v has three edges e_1, e_2, e_3 (possibly parallel) on short paths to $[\kappa]$ to vertices $i_1, i_2, i_3 \in [\kappa]$ (not necessarily distinct). Then there is a minimal structure S which contains $v, e_1, e_2, e_3, i_1, i_2, i_3$, and a short path from v to $[\kappa]$ via each edge e_1, e_2, e_3. Since S is minimal, there are $0 \leq r \leq 3(\omega - 1)$ light vertices $a_1, \ldots, a_r$ in S which form the short paths from v to $[\kappa]$ via e_1, e_2, e_3. Also, since S is minimal, it contains at most 3ω edges. To consider two extremes, for example, S might be three non-intersecting paths, or a single path with e_1, e_2, e_3 being parallel and all other vertices connected by non-parallel edges.

Observe that each vertex a_i has at least two edges, meaning in the application of Proposition 1 it incurs a fraction $\frac{1}{a_i}$ or less. Applying Proposition 1, the expected number of structures S is asymptotically bounded from above by

$$\kappa^3 M^{3\omega} \sum_{r=0}^{3\omega} \sum_{\kappa < a_1, \ldots, a_r} \frac{1}{v^{3/2}} \prod_{i=1}^{r} \frac{1}{a_i} \lesssim \frac{3\omega\kappa^3 (M \log t)^{3\omega}}{v^{3/2}}.$$

Hence, taken over all $v > \kappa_o$, this is $O\left(\frac{3\omega\kappa^3 (M \log t)^{3\omega}}{\kappa_o^{1/2}}\right) = o(1)$.

The rest of the proof is of a similar nature and is left for the journal version of the paper. □

We define the *truncated r-ball around v*, denoted by $\widetilde{\mathcal{B}}(v, r)$, as follows:

1. Delete from $\mathcal{B}(v, r)$ all edges incident to vertices in $[\kappa]$, denote by $\mathcal{B}_-(v, r)$ the resulting graph.
2. Let $\mathcal{C}_v(\mathcal{B}_-(v, r))$ be the connected component in $\mathcal{B}_-(v, r)$ that contains v. Add to $\mathcal{C}_v(\mathcal{B}_-(v, r))$ all edges (u, v) deleted in the previous step such that $u \in [\kappa]$ and $v \in \mathcal{C}_v(\mathcal{B}_-(v, r))$. The resulting graph is $\widetilde{\mathcal{B}}(v, r)$.

The following is a corollary of the above.

Corollary 1. *With high probability, for every vertex $v > \kappa_o$, $\widetilde{\mathcal{B}}(v, \omega)$ belongs to one of the following categories:*

(i) $\widetilde{\mathcal{B}}(v, \omega)$ *is a tree and all vertices are light.*
(ii) $\widetilde{\mathcal{B}}(v, \omega)$ *has no cycles and one or two heavy vertices.*
(iii) *In $\widetilde{\mathcal{B}}(v, \omega)$, v is part of a short cycle of light vertices, and any heavy vertex in $\widetilde{\mathcal{B}}(v, \omega)$ only connects to v via edges that are part of that cycle.*
(iv) *In $\widetilde{\mathcal{B}}(v, \omega)$, there is a short cycle of light vertices which v is not part of, which connects to v through a short path P, and there is at most one edge e on a path from from v to a heavy vertex in $\widetilde{\mathcal{B}}(v, \omega)$ such that e is not part of P.*

Degree of Outer-Core Vertices. For $i \in [t]$ consider the vertex i and the core $[i]$. Immediately after the vertex i is added, the graph under construction at that point, $PA_i(m, \delta)$, has total degree $2mi$, and $D_i(i)$ is a random variable taking integral value between m and $2m$. We may ask, given $D_i(i) = a$, what is the probability that $D_i(t) = a + d$? The question can be framed as one about a Polya urn process in which the urn initially contains a red balls and $2mi - a$ black balls, and the selection process has weighting functions $W_R(k) = k + \delta$ and $W_B(k) = k - (i - 1)\delta$ for red and black balls respectively (see, e.g., [14]).

Notation: $S_i(t)$ is the sum of degrees of vertices in $[i]$ in $PA_t(m, \delta)$. The following was shown in [2]:

Lemma 3. *There is a constant $C(m, \delta)$ that depends only on m and δ, such that for $1 \leq d \leq n \leq m(t - i)$,*

$$\mathbf{Pr}\left(D_i(t) = a + d \mid S_i(t) - 2mi = n, D_i(i) = a\right) \leq C(m, \delta)\frac{1}{d}\left(\frac{Id}{I + n - d}\right)^{a+\delta}e^{-\frac{dI}{I+n}}$$

and

$$\mathbf{Pr}(D_i(t) = a \mid S_i(t) - 2mi = n, D_i(i) = a) \leq \left(\frac{I}{I + n}\right)^{a+\delta},$$

where $I = I(i, m, \delta) = i(2m + \delta) - 1$.

Furthermore, the following was also given in [2]:

Lemma 4. *Suppose $\delta \geq 0$ and for a vertex $i \in [t]$, $i = i(t) \to \infty$. There exists a constant $K_0 > 0$ that depends only on m and δ, such that the following holds for any constant $K > K_0$ and h which is smaller than a constant that depends only on m, δ,*

$$\mathbf{Pr}\left(S_i(t) < \frac{1}{K}\mathbf{E}[S_i(t)]\right) \leq e^{-hi}.$$

We use these lemmas to prove the following:

Lemma 5. *With high probability, for every $i \in [\kappa_o]$, $D_i(t) \geq \left(\frac{t}{\kappa_o}\right)^\gamma\frac{1}{\kappa_o^2}$.*

Proof. Letting $h = \frac{\log \kappa_o}{\kappa_o}$ in Lemma 4, we see that for some constant K, $S_{\kappa_o}(t) \geq Kt^\gamma\kappa_o^{1-\gamma}$. Let $z = z(t) \to \infty$ as $t \to \infty$ to be determined later. Letting $n = Kt^\gamma\kappa_o^{1-\gamma} - 2m\kappa_o$ and applying Lemma 3,

$$\mathbf{Pr}\left(D_{\kappa_o}(t) \leq \frac{n}{\kappa_o z} \mid S_{\kappa_o}(t) - 2m\kappa_o \geq n, D_{\kappa_o}(\kappa_o) = a\right)$$

$$\lesssim \left(\frac{I}{I + n}\right)^{a+\delta} + \sum_{d=1}^{n/(\kappa_o z)}\left(\frac{I}{I + n - d}\right)^{a+\delta}d^{a+\delta-1}e^{-\frac{dI}{I+n}}$$

$$\leq \left(\frac{I}{I + n}\right)^{a+\delta} + \frac{I^{a+\delta}}{(I + n - n/(\kappa_o z))^{a+\delta}}\sum_{d=0}^{n/(\kappa_o z)}d^{a+\delta-1}.$$

Since $\kappa_o \to \infty$ and $z \to \infty$ as $t \to \infty$, we have $n/(\kappa_o z) = o(n)$, so $\frac{1}{(I+n-n/(\kappa_o z))^{a+\delta}} \lesssim \frac{1}{(I+n)^{a+\delta}}$.
Furthermore,

$$\sum_{d=0}^{n/(\kappa_o z)}d^{a+\delta-1} \lesssim \int_0^{n/(\kappa_o z)}x^{a+\delta-1}\,dx \leq \frac{1}{a+\delta}\left(\frac{n}{\kappa_o z}\right)^{a+\delta}.$$

Hence,

$$\mathbf{Pr}\left(D_{\kappa_o}(t) \leq \frac{n}{\kappa_o z} \mid S_{\kappa_o}(t) - 2m\kappa_o \geq n, D_{\kappa_o}(\kappa_o) = a\right)$$

$$\lesssim \left(\frac{I}{I + n}\right)^{a+\delta} + \left(\frac{I}{I + n}\right)^{a+\delta}\left(\frac{n}{\kappa_o z}\right)^{a+\delta}$$

$$\lesssim \left(\frac{I}{I + n}\right)^{a+\delta} + \frac{1}{z^{a+\delta}}.$$

Now, we choose $z(t) = \kappa_o^2$. Then,

$$\left(\frac{I}{I+n}\right)^{a+\delta} = \left(\frac{\kappa_o(2m+\delta-1)}{\kappa_o(2m+\delta-1) + Kt^\gamma \kappa_o^{1-\gamma} - 2m\kappa_o}\right)^{a+\delta} \lesssim \left(\frac{\kappa_o}{t}\right)^{\gamma(a+\delta)} = o\left(\frac{1}{z^{a+\delta}}\right)$$

since $a \geq m \geq 5$ and $\delta \geq 0$.

Thus,

$$\mathbf{Pr}\left(D_{\kappa_o}(t) \leq \frac{n}{\kappa_o z} \mid S_{\kappa_o}(t) - 2m\kappa_o \geq n, D_{\kappa_o}(\kappa_o) = a\right) \lesssim \frac{1}{\kappa_o^{2(m+\delta)}}.$$

A simple coupling argument shows that $D_{\kappa_o}(t)$ is stochastically dominated by $D_i(t)$ for any $i \in [\kappa_o]$. Therefore, taking the union bound over $[\kappa_o]$ we get $\frac{\kappa_o}{\kappa_o^{2(m+\delta)}} = o(1)$ since $a \geq m \geq 5$ and $\delta \geq 0$. $\qquad\square$

5 Convergence of the Majority Dynamics

In this section we show that the system converges to the initial majority opinion and bound the time it takes. Informally, Lemma 6 shows convergence for a tree when the bias away from $1/2$ is large enough, Lemma 7 demonstrates for vertices outside the outer core, it only takes a constant number of steps for the probability of being red to get below the bias threshold that Lemma 6 requires. It also uses the fact that vertices in this range are almost tree-like. The conclusion is that there is a certain contiguous set of steps when all the vertices outside the outer core are blue. Finally, Lemma 8 shows that when this happens, the vertices in the outer core are all blue. Since there is a time step in which all vertices are blue, the graph remains blue thereafter.

For real p and integer $n > 3$ define

$$f(n,p) = \left[\left(1 + \frac{1}{\sqrt{n-1}}\right)2\right]^{\frac{2}{n-3}} 4p(1-p). \tag{1}$$

The following lemma was essentially first proved by the first author in [1]. Due to space restrictions, we give only an informal overview here.

Lemma 6. *Let $T_u(h, d^+)$ be a depth-h tree rooted at u where all non-leaf vertices have degree at least $d^+ \geq 5$ odd. Let $p \in (0, \frac{1}{2})$, $k \geq 5$ and $d = k \wedge d^+$. Suppose at time $\tau = 0$ each vertex of $T_u(h, d^+)$ is assigned red with probability p. Under $\mathcal{MP}^k$ the probability that the root u is red at time step h is at most $\frac{1}{4}(f(d,p))^{\left(\frac{d-1}{2}\right)^h}$.*

Proof (Overview). Suppose instead of $\mathcal{MP}^k$ we had a modified version $\mathcal{MMP}^k$ on the tree in which each vertex other than the root u assumes its parent is red. Under the same sequence of random choices of which neighbours to poll, $\mathcal{MMP}^k$ can only make it more likely that u ends up being red at time step $\tau = h$. It also has the advantage of breaking dependencies between vertices at the same depth in the tree. Denoting $p_\tau(v)$ the probability of vertex v being red at time step τ, we show that under $\mathcal{MMP}^k$, we get $p = p_0(v_h) > p_1(v_{h-1}) > \ldots > p_{h-1}(v_1) > p_h(v_0)$ where v_i is a child of v_{i-1} in the tree and $v_0 = u$. In fact, the sequence of probabilities decays very rapidly, and we find that $p_h(v_0) < \frac{1}{4}\left(f(d,p)\right)^{\left(\frac{d-1}{2}\right)^h}$. $\square$

Lemma 7. *Let $k \geq 5$ be odd and let $d = m \wedge k$ if m is odd and $d = \min\{m - 1, k\}$ if m is even. Let ε be any positive constant, let $\tau_* = B \log_d \log_d t$ where $B = B(d, \varepsilon) = \frac{1+\varepsilon}{\log_d\left(\frac{d-1}{2}\right)}$ and let α^* be the smallest positive solution for x in the equation $\mathbf{Pr}\left(\mathrm{Bin}(d-1, x) \geq \frac{d-1}{2}\right) = x$. If each vertex in $PA_t(m, \delta)$ is red independently with probability $\alpha < \alpha^*$, then $\mathbf{whp}$ under $\mathcal{MP}^k$ every vertex $v \in [t] \setminus [\kappa_o]$ is blue at time steps $\tau = \tau_* + 1, \tau_* + 2$.*

Proof. Let integer $n \geq 2$, $f(x) = \mathbf{Pr}\left(\mathrm{Bin}(2n, x) \geq n\right)$ and $g(x) = f(x) - x$. Observe $g(0) = 0$ and $g(1/2) > 0$. Furthermore, $g'(x) = \binom{2n}{n}nx^{n-1}(1 - x)^n - 1$, whence $g'(0) = -1$. Therefore $g(x)$ has a root x^* in $(0, 1/2)$. Now $g''(x) = \binom{2n}{n}nx^{n-2}(1 - x)^{n-1}[(n - 1) - x(2n - 1)]$ which is strictly positive on $0 < x < \frac{1}{2} - \frac{1}{2(2n-1)}$ and non-positive on $\frac{1}{2} - \frac{1}{2(2n-1)} \leq x < 1$. We can therefore deduce that x^* is the unique root of $g(x)$ in $(0, 1/2)$, and that for the interval $[c_1, c_2]$ where $0 < c_1 < c_2 < x^*$, g attains a maximum at c_1 or c_2. Hence, for a given $x \in [c_1, c_2]$, we have $0 < f(x) < x$ and $x - f(x) = -g(x) > -\max\{g(c_1), g(c_2)\} > 0$. Therefore, we need only a constant number of iterations of f until $f(f(\ldots f(x))\ldots) < c_1$. When $2n = d - 1$, we write $\alpha^* = x^*$.

Now consider a rooted tree of depth h where non-leave vertices have $2n = d - 1$ children, and suppose that each vertex is coloured red independently with probability $\alpha < \alpha^*$ at time $\tau = 0$. By the same argument as in the proof of Lemma 6, at time $\tau = 1$ the depth $h - 1$ vertices are red independently with probability $f(\alpha) < \alpha - c_2$. Continuing in this way, the probability the root is red is at most c_1 if $h > c_3$ where c_3 is a large enough finite constant.

Let $\tau' = \tau_* - c_3$ and suppose that $\mathcal{B}(v, \omega)$ is a tree. Since $\omega = A \log \log t$ with A arbitrarily large, then we may assume $\omega = a \log_d \log_d t$ where a is a constant such that $\omega \geq \tau_* + 3$. This means $\mathcal{B}(w, \tau_*)$ is a tree if w is a neighbour of v or v itself. By the above, we may therefore assume that at time $t = c_3$, the depth $\tau_* - c_3 = \tau'$ vertices are red independently with probability at most c_1.

Then by Lemma 6 the probability v is red at time step τ_* is at most $\frac{1}{4}f(d, c_1)^{\left(\frac{d-1}{2}\right)^{\tau'}}$. For large enough t, $\tau' \geq \frac{1+\varepsilon/2}{\log_d\left(\frac{d-1}{2}\right)} \log_d \log_d t$, therefore

$$\left(\frac{d-1}{2}\right)^{\tau'} \geq \left(\frac{d-1}{2}\right)^{\frac{1+\varepsilon/2}{\log_d\left(\frac{d-1}{2}\right)} \log_d \log_d t} = d^{(1+\varepsilon/2)\log_d \log_d t} = (\log_d t)^{1+\varepsilon/2}.$$

Thus,

$$f(d, c_1)^{\left(\frac{d-1}{2}\right)^{\tau'}} \leq d^{-\log_d\left(\frac{1}{f(d,c_1)}\right)(\log_d t)^{1+\varepsilon/2}} = t^{-\log_d\left(\frac{1}{f(d,c_1)}\right)(\log_d t)^{\varepsilon/2}}.$$

If $f(d, c_1) < \beta < 1$ where β is a constant then the above is at most $t^{-(\log_d t)^{\varepsilon/4}}$ when t is large enough. By the same logic, and since each of the children of v

are also trees out to distance τ_*, the same probability bound applies to them. Thus, taking the union bound, we see that all vertices v such that $\mathcal{B}(v, \omega)$ is a tree are blue at times $\tau_*, \tau_* + 1, \tau_* + 2$.

We extend the above to other vertices outside $[\kappa_o]$. From Corollary 1, we see that v always has at most two "bad" edges that it can assume are always red. Since $m \geq 5$, this leaves $m - 2 \geq 3$ "good" edges which, if they are blue, will out-vote the bad edges, regardless of what their actual colours are. Thus, suppose $e_1 = (v, w_1), \ldots, e_{m-2} = (v, w_{m-2})$ are good edges. As per the proof of Lemma 6, the random variables $Y_\tau(w_i)$ for $i \in \{1, \ldots, m - 2\}$ depend only on vertices in the subtree of $\widetilde{\mathcal{B}}(v, \omega)$ for which w_i is a root. This is a depth–$(\omega - 1)$ tree where each vertex not a leaf nor root has at least $m - 1$ children. Since we may assume that $\omega \geq \tau_* + 2$, it follows by the above that **whp**, all such w_i are blue at time steps $\tau_*, \tau_* + 1, \tau_* + 2$. This forces v to be blue at time steps $\tau_* + 1, \tau_* + 2, \tau_* + 3$. Thus, we have proved that **whp**, all vertices $v \in [t] \setminus [\kappa_o]$ are blue at time steps $\tau_* + 1, \tau_* + 2$. $\qquad\square$

It remains to consider the vertices in $[\kappa_o]$:

Lemma 8. *If every vertex in $v \in [t] \setminus [\kappa_o]$ is blue at time step $\tau_* + 1$, then* **whp** *every $v \in [\kappa_o]$ is blue at time step $\tau_* + 2$.*

Proof. Consider a vertex $v \in [\kappa_o]$. We partition v's set of incident edges E_v in $PA_t(m, \delta)$ into two sets $E_{v1} = \{(v, w) : w \in [\kappa_o]\}$ and $E_{v2} = E_v \setminus E_{v1}$. Clearly, $|E_{v1}| \leq m\kappa_o$, so by Lemma 5, we may assume that $|E_{v2}| \geq \left(\frac{t}{\kappa_o}\right)^\gamma \frac{1}{\kappa_o^2} - m\kappa_o$ for every $v \in [\kappa_o]$. Consequently, the probability that at time step $\tau_* + 1$ the majority of edges picked by v are in E_{v1} is zero if $d \geq 2|E_{v1}| + 1$ and $O\left(\mathbf{Pr}\left(\text{Bin}(d, \frac{\kappa_o^4}{t}) > \frac{d}{2}\right)\right) = O\left(\kappa_o^4/t\right)$ if $d \leq 2|E_{v1}|$. Taken over all vertices in $[\kappa_o]$ this is $o(1)$. $\qquad\square$

Corollary 2. *With high probability, $PA_t(m, \delta)$ is entirely blue at all time steps $\tau \geq \tau_* + 2$.*

6   Conclusion and Open Problems

We have seen that with high probability, local majority dynamics on preferential attachment graphs with power law exponent at least 3 very rapidly converges to the initial majority when the initial distribution of red vs. blue opinions is sufficiently biased away from equality. The speed of convergence is affected both by the number of neighbours polled at each step as well structural parameters of the graph, specifically, how many edges are added when a new vertex joins in the construction process of the graph.

A natural next step would be to analyse the process for $-m < \delta < 0$, which generates graphs with power-law exponents between 2 and 3. These appear to better reflect "real world" networks, but our experience suggests that structural differences make the techniques of this paper ineffective in this regime.

Another direction would be to explore how adversarial placements of opinions affects outcome, as studied in [8] for random regular graphs.

References

1. Abdullah, M.A., Draief, M.: Global majority consensus by local majority polling on graphs of a given degree sequence. Discrete Appl. Math. **180**, 1–10 (2015)
2. Abdullah, M.A., Fountoulakis, N.: A phase transition in the evolution of bootstrap percolation processes on preferential attachment graphs, arXiv:1404.4070
3. Albert, R., Barabási, A.-L.: Statistical mechanics of complex networks. Rev. Mod. Phys. **74**, 47–97 (2002)
4. Aldous, D., Fill, J.: Reversible Markov Chains and Random Walks on Graphs, (in preparation) http://stat-www.berkeley.edu/pub/users/aldous/RWG/book.html
5. Bollobás, B., Riordan, O.: The diameter of a scale-free random graph. Combinatorica **24**, 5–34 (2004)
6. Bollobás, B., Riordan, O., Spencer, J., Tusnády, G.: The degree sequence of a scale-free random graph process. Random Struct. Algorithms **18**, 279–290 (2001)
7. Buckley, P.G., Osthus, D.: Popularity based random graph models leading to a scale-free degree sequence. Discrete Math. **282**, 53–68 (2004)
8. Cooper, C., Elsässer, R., Radzik, T.: The power of two choices in distributed voting. In: Esparza, J., Fraigniaud, P., Husfeldt, T., Koutsoupias, E. (eds.) ICALP 2014, Part II. LNCS, vol. 8573, pp. 435–446. Springer, Heidelberg (2014)
9. Cooper, C., Frieze, A.: The cover time of the preferential attachment graph. J. Comb. Theor. Ser. B **97**, 269–290 (2004)
10. Cruise, J., Ganesh, A.: Probabilistic consensus via polling and majority rules. In: Proceeidngs of Allerton Conference (2010)
11. Dorogovtsev, S.N., Mendes, J.F.F., Samukhin, A.N.: Structure of growing networks with preferential linking. Phys. Rev. Lett. **85**, 4633–4636 (2000)
12. Drinea, E., Enachescu, M., Mitzenmacher, M.: Variations on random graph models for the web. Technical report TR-06-01, Harvard University, Department of Computer Science (2001)
13. Hassin, Y., Peleg, D.: Distributed probabilistic polling and applications to proportionate agreement. Inf. Comput. **171**, 248–268 (2001)
14. van der Hofstad, R.: Random Graphs and Complex Networks (2013). http://www.win.tue.nl/~rhofstad/NotesRGCN.pdf
15. Mossel, E., Neeman, J., Tamuz, O.: Majority dynamics and aggregation of nformation in social networks (2012). arXiv:1207.0893
16. Simon, H.A.: On a class of skew distribution functions. Biometrika **42**, 425–440 (1955)
17. Yule, G.U.: A mathematical theory of evolution, based on the conclusions of Dr. J.G. Willis F.R.S. Phil. Trans. Roy. Soc. Lond. B **213**, 21–87 (1925)

Rumours Spread Slowly in a Small World Spatial Network

Jeannette Janssen[1] and Abbas Mehrabian[2,3(✉)]

[1] Department of Mathematics and Statistics,
Dalhousie University, Halifax, NS B3H 3J5, Canada
[2] University of Waterloo, Waterloo, ON, Canada
amehrabi@uwaterloo.ca
[3] Pacific Institute for Mathematical Sciences, Vancouver, BC, Canada

Abstract. Rumour spreading is a protocol that models the spread of information through a network via user-to-user interaction. The spread time of a graph is the number of rounds needed to spread the rumour to the entire graph. The Spatial Preferred Attachment (SPA) model is a random graph model that models complex networks: vertices are placed in a metric space, and the link probability depends on the metric distance between vertices, and on their degree. We show that the SPA model typically produces graphs that have small effective diameter, i.e. $O(\log^3 n)$, while rumour spreading is relatively slow, namely polynomial in n.

1 Introduction

There is increasing consensus in the scientific community that complex networks (e.g. on-line social networks or citation graphs) can be accurately modelled by spatial random graph models. Spatial random graph models are models where the vertices are located in a metric space, and links are more likely to occur between vertices that are close together in this space. The space can be interpreted as a *feature space*, which models the underlying characteristics of the entities represented by the vertices. Specifically, entities with similar characteristics (for example, users in a social network that share similar interests) will be placed close together in the feature space. Thus the distance between vertices is a measure of affinity, and thus affects the likelihood of the occurrence of a link between these vertices.

An important reason to model real-life networks is to be able, through simulation or theoretical analysis, to study the dynamics of information flow through the network. Several ways to model flow of information through a network have

J. Janssen—The collaboration between the authors was a result of the visit of the first author to the Institute of Mathematics and Applications (IMA) in Minnesota. She wishes to thank IMA for providing this opportunity. She also acknowledges NSERC for their support of this research.

A. Mehrabian—Supported by the Vanier Canada Graduate Scholarships program and a PIMS Postdoctoral Fellowship.

© Springer International Publishing Switzerland 2015
D.F. Gleich et al. (Eds.): WAW 2015, LNCS 9479, pp. 107–118, 2015.
DOI: 10.1007/978-3-319-26784-5_9

been proposed recently, based on metaphors such as the spread of infection or of fire, or the range of a random walk through the graph [5,7,21,23]. Here we focus on a protocol called *rumour spreading*. It differs from the models based on fire or infection in that in each round, the rumour spreads to only one neighbour of each informed vertex. On the other hand, the difference with a random walk approach is that each informed vertex spreads the rumour, and thus we have more of a growing tree of random walks.

In this paper we study the behaviour of the rumour spreading protocol on graphs produced by the Spatial Preferential Attachment (SPA) model. The SPA model is a spatial model that produces sparse power law graphs. We show that, on the one hand, the graph distance between vertices in such a graph tends to be small (polylogarithmic in n, the number of vertices), while on the other hand, it takes a long time (polynomial in n) to spread the rumour to most of the vertices.

1.1   The SPA Model

The SPA model is a growing graph model, where one new vertex is added to the graph in each time step. The vertices are chosen from a metric space. Each vertex has a sphere of influence, whose size grows with the degree of the vertex. A new vertex can only link to an existing vertex if it falls inside its sphere of influence. Therefore, links between vertices depend on their (spatial) distance, and on the in-degree of the older vertex.

The SPA model was introduced in [2], where it was shown that asymptotically, graphs produced by the SPA model have a power law degree distribution with exponent in $[2, \infty)$ depending on the parameters. The model was further studied in [8,18,19]. The model can be seen as a special case of the spatial model introduced by Jacob and Mörters in [17] and further studied in [16]. The SPA model has similarities with the spatial models presented in [4,6,12,25].

Let S be the unit hypercube in $\mathbb{R}^m$, equipped with the torus metric derived from the Euclidean norm. The SPA model stochastically generates a graph sequence $\{G_t\}_{t \geq 0}$; for each $t \geq 0$, $G_t = (V_t, E_t)$, where E_t is an edge set, and $V_t \subseteq S$ is a vertex set. The index t is an indication of time. The in-degree, out-degree and total degree of a vertex v at time t is denoted by $\deg^-(v,t)$, $\deg^+(v,t)$ and $\deg(v,t)$, respectively.

We now define the *sphere of influence* $S(v,t)$ of vertex v at time t. Let

$$A(v,t) = \frac{A_1 \deg^-(v,t) + A_2}{t},$$

where $A_1, A_2 > 0$ are given parameters. If $A(v,t) \leq 1$, then $S(v,t)$ is defined as the ball, centred at v, with total volume $A(v,t)$. If $A(v,t) > 1$ then $S(v,t) = S$, and so $|S(v,t)| = 1$. To keep the second option from happening often, we impose the additional restriction that $A_1 < 1$; this ensures that in the long run, $S(v,t) \ll 1$ for all v.

The generation of a SPA model graph begins at time $t = 0$ with G_0 being the null graph. At each time step $t \geq 1$, a node v_t is chosen from S according to the uniform distribution, and added to V_{t-1} to form V_t. Next, independently for each vertex $u \in V_{t-1}$ such that $v_t \in S(u,t)$, a directed link (v_t, u) is created with probability p.

Because the volume of the sphere of influence of a vertex is proportional to its in-degree, so is the probability of the vertex receiving a new link at a given time. Thus link formation is governed by a preferential attachment, or "rich get richer", principle, which leads to a power law degree distribution of the in-degrees, and thus also of sizes of the spheres of influence.

Another important feature of the model is that all spheres of influence tend to shrink over time. This means that the length of an edge (the distance between its endpoints) depends on the time when it was formed: edges formed in the beginning of the process tend to be much longer than those formed later (see [18] for more on the distribution of edge lengths). As we will see, the old, long links significantly decrease the graph distance between vertices. This is a feature unique to the SPA model; "static" variations of the SPA model such as that presented in [3], tend to limit the maximum length of an edge, which leads to a larger diameter.

Note that the SPA model generates directed graphs. However, the rumour spreading protocols we study here completely ignore the edge orientations; we imagine that they work on the corresponding undirected underlying graph. Similarly, in estimating the graph distances, we ignore the edge orientations.

1.2 Rumour Spreading

Rumour spreading is a model for the spread of one piece of information, the *rumour*, which starts at one vertex, and in each time step, spreads along the edges of the graph according to one of the following protocols.

The *push protocol* is a round-robin rumour spreading protocol defined as follows: initially one vertex of a simple undirected graph knows a rumour and wants to spread it to all other vertices. In each round, every informed vertex sends the rumour to a random neighbour.

The *push&pull protocol* is another round-robin rumour spreading protocol defined as follows: initially one vertex of a simple undirected graph knows a rumour and wants to spread it to all other vertices. In each round, every informed vertex sends the rumour to a random neighbour, while every uninformed vertex contacts a random neighbour and gets the rumour from her if she knows it.

In both protocols defined above, all vertices work in parallel. These are synchronized protocols, so if a vertex receives the rumour at round t, it starts passing it on from round $t + 1$. Also, vertices do not have memory, so a vertex might contact the same neighbour in consecutive rounds.

We are interested in the *spread time*, the number of rounds needed for all vertices to get informed. Since the SPA model does not generally produce connected graphs, we here limit this requirement to vertices in the same component as the starting vertex. It is clear that the push&pull protocol is generally quicker.

The push protocol was defined in [14] for the complete graph, and was studied in [11] for general graphs. The push&pull protocol was defined in [9], where experimental results were presented, and the first analytical results appeared in [20]. For a summary of results on the push&pull protocol, see [1, Table 1].

1.3   Main Results

Clearly, the diameter of a graph is a lower bound on the spread time, at least for appropriate choices of starting vertex. An easy well known upper bound for spread time is $O(\Delta(\text{diameter} + \log n))$ [11, Theorem 2.2], where Δ denotes the maximum degree, so in graphs of bounded degree, spreading time is largely determined by the diameter. Another important factor in spreading time is the degree distribution of the graph. Vertices of high degree tend to slow down the spread, since only one neighbour of a vertex is contacted in each round. SPA model graphs have a power law degree distribution, and the maximum degree is typically $\Omega(n^{A_1})$ (see [2]).

In this paper we prove two main results. First, we show that for most pairs of vertices, the graph distance is polylogarithmic in the number of vertices. Thus, SPA model graphs are so-called small worlds. SPA model graphs are generally not connected, and the size and threshold of the giant component are not exactly known. Therefore we state our result in terms of the *effective diameter*, introduced in [22]. A graph G has effective diameter at most d if, for at least 90 % of all pairs of vertices of G that are connected, their graph distance is at most d. We say an event happens asymptotically almost surely (a.a.s.) if its probability approaches 1 as n goes to infinity. All logarithms are in the natural base in this paper.

Recall that the SPA model has four parameters: $m \in \mathbb{Z}_+$ is the dimension, $A_1, A_2 > 0$ control the volumes of vertices' spheres of influence, and $p \in (0, 1]$ is the probability of link formation.

Theorem 1. *For each choice of $A_1 \in [0, 1)$, and for large enough choice of A_2, a.a.s. a graph produced by the SPA model with parameters $A_1, A_2, p = 1$ and $m = 2$ has effective diameter $O(\log^3 n)$.*

Remark 1. The constant 90 % in the definition of effective diameter is somewhat arbitrary. Our arguments yield similar bounds if this is changed to any other constant strictly smaller than 100 %.

As noted before, this result refers to the *undirected* diameter. In [8], it was shown that a.a.s. any shortest directed path has length $O(\log n)$. This result does not apply to our situation, since pairs connected by a directed path are a small minority.

Our second result illustrates that, in spite of the small world property, a.a.s. rumour spreading with the push protocol is slow, that is, takes polynomial time in n (a polynomial lower bound for the push&pull protocol is also given in Corollary 1).

Theorem 2. *Let G be a graph produced by the SPA model with parameters $A_1, A_2 > 0, m = 2$, and assume $pA_1 < 1$. Let $\alpha > 0$ be so that*

$$\alpha < \min\{pA_1/24, (1 - pA_1)/12\} \in (0, 1).$$

If a rumour starts in G from a uniformly random vertex, then a.a.s. after n^α rounds of the push protocol, the number of informed vertices is $o(n)$.

Remark 2. The main objective of this theorem (and Corollary 1) is to give *some* polynomial lower bound for the number of rounds needed to inform $\Omega(n)$ vertices. In particular, we have not tried to optimize the exponent α here.

We can understand this result as follows. While SPA model graphs have a backbone of long edges that decrease graph distances between vertices, only old edges are long. Old edges have old endpoints, so the vertices on this backbone are old. Old vertices have high degree, and vertices of high degree are slower in spreading the rumour. So if the rumour travels along long edges, then it will become delayed due to high vertex degree, and if it travels along short edges, it takes many steps to cover the entire space.

In [17] it was shown that, for certain choices of the parameters, the generalized spatial model by Jacob and Mörters exhibits a similar mixture of long and short edges. This suggests that our results may be extended to this model; this would be an interesting question to pursue.

The push&pull protocol has been studied on two small-world (non-spatial) models and it turned out that it spreads the rumour in logarithmic time: it was shown in [10] that on a random graph model based on preferential attachment, push&pull spreads the rumour within $O(\log n)$ rounds. A similar bound was proved in [13] for the performance of this protocol on random graphs with given expected degrees when the average degree distribution is power law. Thus, the SPA model is a unique example of a natural model that exhibits both the small world property and slow rumour spreading.

2   The Effective Diameter of SPA Model Graphs

In this section we show that a SPA model graph typically has a small effective diameter. We assume that S is two-dimensional, so $m = 2$, and $p = 1$. We will derive our bound using properties of the random geometric graph model, especially those studied in [15,24].

A two-dimensional random geometric graph on N vertices with radius $r = r(N)$ (denoted by $RGG(N,r)$) is generated as follows: N vertices are chosen independently and uniformly at random from the unit square S, and an edge is added between two vertices if and only if their Euclidean distance is at most r. To see how the geometric random graph model relates to the SPA model, let $\{G_t\}_{t=0}^n$ be a sequence of graphs produced by the SPA model. For each t, define the graph R_t as a graph with vertex set $V(G_t)$ in which two vertices are adjacent if and only if their distance is at most $\sqrt{A_2/(t\pi)}$. Observe that R_t conforms to the random geometric graph model on t vertices with radius

$$r_t := \sqrt{A_2/(t\pi)}.$$

Moreover, observe that for all t, R_t is a subgraph of (the undirected underlying graph of) G_t. For, at all times from 1 to t each sphere of influence has volume at least A_2/t, i.e. radius at least r_t. Therefore, if two vertices v_i and v_j, $1 \leq i < j \leq t$, have distance at most r_t, then at time j, when v_j is born, v_j will fall inside the sphere of influence of v_i, and a link $v_j v_i$ will be created. We will use the graphs R_t to bound the diameter of G_n.

As mentioned earlier, graphs produced by the SPA model are generally not connected. However, we can choose the parameters so that there exists a giant component, i.e. a component containing an $\Omega(1)$ fraction of all vertices. Note that if R_n has a giant component, then so has G_n. Moreover, it is known (see [24]) that there exists a constant a_c so that, if $r = \sqrt{a/(\pi N)}$ with $a > a_c$, then a.a.s. $RGG(N, r)$ has a giant component, while if $a < a_c$ then a.a.s. it does not have one (note that a is simply the average degree). Experiments give $a_c \approx 4.51$. Therefore, G_n has a giant component a.a.s. if $A_2 > a_c$. It would be interesting to determine whether this value of A_2 is indeed the threshold for the emergence of the giant component in G_n. Determination of this threshold was left open in [8].

The following theorem about the size of the giant component directly follows from Theorem 10.9 and Proposition 9.21 in [24].

Theorem 3 [24]. *There exists a constant $a_{big} > a_c$ so that a.a.s. a random geometric graph $RGG(N, r)$ with $\pi N r^2 > a_{big}$ has a connected component containing at least $0.99N$ of its vertices.*

Lemma 1. *Let $G = G_n$ be a graph produced by the SPA model with parameters $A_1 \in [0, 1)$, $A_2 > a_{big}$, $m = 2, p = 1$. Let C be the giant component of G_n. Then a.a.s. for every t with $\log n \leq t \leq n$, the giant components of R_t and $R_{t/2}$ intersect.*

Proof. For each $\log n \leq t \leq n$, let C_t be the giant component of R_t. Since $A_2 > a_{big}$, a.a.s. each C_t contains at least $0.99t$ vertices. Therefore, for all t, C_t and $C_{t/2}$ intersect. □

We want to show that the diameter of the giant component of G_n is $O(\log^3 n)$. We will proceed as follows. Given an arbitrary vertex v in the giant component of G_n, the idea is to find a path of length $O(\log^2 n)$ connecting v to some vertex y_1 in the giant component of $G_{n/2}$, then connect y_1 to a vertex y_2 in the giant of $G_{n/4}$ etc.

We will use a known structural property of the giant component of a random geometric graph. Suppose R is a random geometric graph with parameters N and $r = r(N)$, and assume that S is subdivided into subsquares of size $r/4 \times r/4$. Two subsquares are called *adjacent* if they share a side. A *path of subsquares* is a sequence of distinct subsquares so that each consecutive pair is adjacent. A path is *closed* if its first and last subsquares are adjacent. A subsquare is called *occupied* if it contains a vertex of R, and *empty* otherwise.

Lemma 2. *There exists an absolute (large) constant M with the following property. Let R be a random geometric graph $RGG(N, r)$ with $\pi N r^2 = M$. Then for large N, with probability at least $1 - N^{-2}$, R has a unique giant component, and every point of S is enclosed by a closed path P of occupied subsquares with the following properties:*

$Mr \log N$

$Mr \log N$

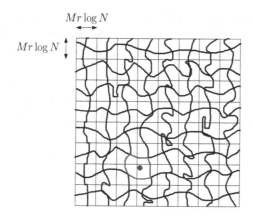

Fig. 1. Illustration of the proof of Lemma 2: left-to-right crossings are blue, top-to-bottom crossings are green. The green and blue crossings constitute a grid-like backbone for the giant component. Any point in the unit square (purple) is enclosed by some closed path P (red) (Color figure online).

(i) all subsquares of P lie inside a $2Mr \log N \times 2Mr \log N$ square, and
(ii) all vertices inside the subsquares of P belong to the giant component.

Proof. For a given rectangle L, a *left-to-right crossing* is a path of occupied subsquares inside L with one endpoint touching the left side of L and the other endpoint touching its right side. A *top-to-bottom crossing* is defined similarly. Partition the unit square S into $(Mr \log N)^{-1}$ horizontal rectangles of size $Mr \log N \times 1$, and also into $(Mr \log N)^{-1}$ vertical rectangles of size $1 \times Mr \log N$. By [15, Lemma 2], each of the horizontal (vertical) rectangles has a left-to-right (top-to-bottom) crossing with probability at least $1 - N^{1/2 - \Omega(M^2)}$. The total number of rectangles is less than $\sqrt{N}$, so by the union bound we can choose M large enough that with probability at least $1 - N^{-2}$ all horizontal (vertical) rectangles have a left-to-right (top-to-bottom) crossing (see Fig. 1). These crossings constitute a grid-like "backbone" for the giant component. Any point of S not part of the giant component lies inside a "cell" of this grid, which lies inside a $2Mr \log N \times 2Mr \log N$ square, completing the proof.   □

Lemma 3. *Let G be a graph produced by the SPA model with parameters $A_1 \in [0, 1)$, $A_2 > a_{big}$, $m = 2$, and $p = 1$. The following is true a.a.s. Let t be an integer such that $\log n \leq t \leq n/2$. Let v be a vertex in the giant component of R_{2t}. Then there exists a path of length $O(\log^2 t)$ in R_{2t} from v to the giant component of R_t.*

Proof. Recall that R_t is a random geometric graph on t vertices with radius $r_t = \sqrt{A_2/(\pi t)}$. Assume that S is subdivided into subsquares of size $r_t/4 \times r_t/4$. By Lemma 2 (applied to R_t), with probability at least $1 - t^{-2}$ there exists a giant component C_t and a constant M with the properties given in that lemma. Assume this is the case (and to obtain the result for all $\log n \leq t \leq n/2$, we simply apply the union bound).

Fix v in the giant component C_{2t} of R_{2t}. If $v \in V(C_t)$ we are done; otherwise, by Lemma 2 there exist a closed path P of subsquares with properties (i) and (ii). Let Q denote the region enclosed by P. By Lemma 1, C_t and C_{2t} intersect, so there exists a shortest path ξ in R_{2t} from v to $V(C_t)$. It can be shown using a geometric argument that ξ can have at most two vertices outside Q. Also, since ξ is a geodesic path, it can contain at most one vertex from each subsquare. Since there are at most $4M^2 \log^2 t$ subsquares in Q, the conclusion follows. □

Theorem 4. *Let $G = G_n$ be a graph produced by the SPA model with parameters $A_1 \in [0,1)$, $A_2 > a_{big}, m = 2, p = 1$. Then a.a.s. R_n has a giant component C_n which contains at least $\sqrt{0.9n}$ vertices, and the diameter of C_n is $O(\log^3 n)$. So the effective diameter of G_n is $O(\log^3 n)$.*

Proof. By Theorem 3 and the fact that R_n is distributed as $RGG(n, r_n)$ with $r_n = \sqrt{A_2/(\pi n)}$, we have that a.a.s. R_n contains a giant component C_n which contains at least $0.99n > \sqrt{0.9n}$ vertices. Assume this is the case. Let y_0 be a vertex in C_n. By repeated application of Lemma 3, a.a.s. there exists a sequence of vertices $y_1, y_2, \ldots, y_k$, where k is the largest integer so that $n2^{-k} \geq \log n$, with the following properties: for each i, $0 < i \leq k$, y_i is in the giant component of $R_{n2^{-i}}$, and there exists a path of length $O(\log^2 n)$ from y_i to y_{i-1}. As $R_{n2^{-k}}$ has size at most $2 \log n$, there exists a path of length $O(\log^3 n)$ between any two vertices in C_n. □

Theorem 1 follows immediately from the above theorem. The methods used in this section do not suffice to show that the diameter of the giant component of G_n is also logarithmic. In principle, it could be that there exist vertices in G_n that are not contained in the giant component of R_n, but that are connected to this component by a long path that uses edges from inside the minor components of the graphs $R_n, R_{n/2}$, etc. Nevertheless, we believe that the SPA model graphs have logarithmic diameters a.a.s., and we leave this as an open problem.

3   Lower Bounds for Rumour Spreading

We will first establish some structural properties of the SPA model, and then use them to prove results about the rumour spreading protocols. Recall that m denotes the dimension and n denotes the number of vertices. Let c_m denote the volume of the m-dimensional ball of unit radius, and let $w = w(n)$, $L = L(n)$, $y = y(n)$ and $T = T(n)$ be functions satisfying

$$c_m L^m > w^{pA_1-1} \log^2 n, \quad y/w = n^{\Omega(1)}, \quad \text{and} \quad y < n/\log n. \tag{1}$$

We say an edge is *long* if the distance between its endpoints is larger than L, and is *short* otherwise. A vertex/edge is *old* if it was born during one of the rounds $1, 2, \ldots, w$, and is *new* otherwise. The following lemma, whose proof is somewhat technical and is omitted from this extended abstract, establishes properties of old and new vertices and long and short edges.

Lemma 4 (Structural properties of graphs generated by the SPA model). *Let G be a graph generated by the SPA model, and let ζ be a uniformly random vertex of it. A.a.s. we have the following properties.*

(a) All new edges are short.
(b) For all $t = 1, 2, \ldots, T$, The number of old vertices within distance tL of ζ is $O\left(\log T + w(tL)^m\right)$.
(c) If there exists a constant $\phi \in (0, 1)$ such that

$$(TL)^m w \left((w/y)^{\phi p A_1/2} + n^{-4/3} \log n\right) = o(1), \tag{2}$$

then all old vertices v within distance TL of ζ satisfy

$$\frac{\deg(v, w)}{\deg(v, n)} < n^\varepsilon \left(\frac{y \log n}{n}\right)^{p A_1}. \tag{3}$$

(d) If there exists a constant $\theta \in (0, 1)$ such that

$$w \left((w/y)^{\theta p A_1/2} + n^{-4/3} \log n\right) = o(1), \tag{4}$$

then all old vertices v satisfy (3).

Since by part (a) of the lemma all edges created after round w are short, assertion (3) quantifies the informal statement "most edges incident to an old vertex are short."

Theorem 5 (Main Theorem for the push protocol). *Suppose that (1) and (2) hold, and suppose there exists a constant $\varepsilon \in (0, 1)$ such that*

$$n^{\varepsilon - p A_1} (y \log n)^{p A_1} \left(T \log T + w L^m T^{m+1}\right) = o(1). \tag{5}$$

Then, if the rumour starts from a uniformly random vertex, then a.a.s. after T rounds of the push protocol, all informed vertices lie within distance TL of the initial vertex.

Proof. Let G be a graph generated by the SPA model, and let ζ be a random vertex. Assume that G and ζ satisfy properties (a)–(c) given in Lemma 4. We need only show that, a.a.s. the rumour does not pass through a long edge during the first T rounds. Note that new vertices are not incident to long edges by Lemma 4(a). Moreover, by Lemma 4(c) every old vertex v within distance TL of ζ satisfies (3), which guarantees that most edges incident to v are short. Condition (3) implies that the probability that an informed old vertex pushes the rumour along a long edge in a given round is smaller than $n^\varepsilon (y \log n/n)^{p A_1}$. On the other hand, for any $t \in \{1, \ldots, T\}$, after t rounds, the number of informed old vertices is $O(\log T + w(tL)^m)$ by Lemma 4(b) and an inductive argument. We apply the union bound over all rounds $t = 1, \ldots, T$:

$$\sum_{t=1}^{T} (\log T + w(tL)^m) \, n^\varepsilon \left(\frac{y \log n}{n}\right)^{p A_1}$$

$$\leq n^{\varepsilon - p A_1} (y \log n)^{p A_1} \left(T \log T + w L^m T^{m+1}\right) = o(1).$$

$\square$

Proof (of Theorem 2). Define $\iota := \min\{pA_1/24, (1 - pA_1)/12\} \in (0, 1)$ and set $w = n^{1/6}$, $y = n^{2/3}$, $L = n^{(\iota-\alpha)/2+(pA_1-1)/12}$, and $T = n^\alpha$, and observe that (1), (2) and (5) hold. By Theorem 5, a.a.s. after T rounds of push protocol, all informed vertices lie in a disc of area $\pi(TL)^2$. By the Chernoff bound, a.a.s. there are $O(n(TL)^2) = o(n)$ vertices in this disc.

Theorem 6 (Main Theorem for the push&pull protocol). *Suppose* (1) *and* (4) *hold, and suppose there exists a constant* $\varepsilon \in (0, 1)$ *with*

$$Twn^{\varepsilon-pA_1}(y \log n)^{pA_1} = o(1). \tag{6}$$

Then, if the rumour starts from a uniformly random vertex, then a.a.s. after T rounds of the push&pull protocol, all informed vertices lie within distance TL of the initial vertex.

Proof. Let G be a graph generated by the SPA model, and let ζ be a uniformly random vertex of it. We may assume that G and ζ satisfy the properties (a) and (c) given in Lemma 4. We need only show that the rumour does not pass through a long edge during the first T rounds. Note that new vertices are not incident to long edges by Lemma 4(a). Moreover, by Lemma 4(c) every old vertex v satisfies (3), which guarantees that most edges incident to v are short. Condition (3) implies that the probability that an old vertex contacts a neighbour along some long edge in a given round is smaller than $n^\varepsilon (y \log n/n)^{pA_1}$. There are exactly w old vertices. By the union bound over all old vertices and over the rounds 1 to T, the probability that an old vertex contacts a neighbour along some long edge is bounded by $wTn^\varepsilon (y \log n/n)^{pA_1}$, which is $o(1)$ by (6).   □

Corollary 1. *Let $\delta > 0$ be an arbitrarily small constant and assume that $m = 2$ and $pA_1 < 1$. Define*

$$\lambda := \frac{pA_1(1 - pA_1)}{10 + 2pA_1} \in (0, 1).$$

If the rumour starts from a uniformly random vertex, then a.a.s. after $n^{\lambda-2\delta}$ rounds of push&pull protocol, number of informed vertices is $o(n)$.

Proof. Define $\mu := pA_1$ and set $w = n^{\mu/(5+\mu)}$, $y = n^{(3+\mu)/(5+\mu)}$, $L = n^{\delta+\mu(\mu-1)/(2\mu+10)}$, and $T = n^{\lambda-2\delta} = n^{\mu(1-\mu)/(10+2\mu)-2\delta}$, and observe that (1) and (4) and (6) are satisfied. By Theorem 6, a.a.s. after T rounds all informed vertices lie in a disc of area $\pi(TL)^2$. By the Chernoff bound, a.a.s. the number of vertices in any such disc is $O(n(TL)^2) = o(n)$.   □

References

1. Acan, H., Collevecchio, A., Mehrabian, A., Wormald, N.: On the push&pull protocol for rumour spreading (2015). Conference version in PODC 2015. arXiv:1411.0948
2. Aiello, W., Bonato, A., Cooper, C., Janssen, J., Prałat, P.: A spatial web graph model with local influence regions. Internet Math. **5**(1–2), 175–196 (2008)

3. Bonato, A., Gleich, D., Mitsche, D., Prałat, P., Tian, Y., Young, D.: Dimensionality of social networks using motifs and eigenvalues. PLoS ONE **9**(9), e106052 (2014)
4. Bonato, A., Janssen, J., Prałat, P.: Geometric protean graphs. Internet Math. **8**(1–2), 2–28 (2012)
5. Bonato, A., Janssen, J., Roshanbin, E.: Burning a graph as a model of social contagion. In: Bonato, A., Graham, F.C., Prałat, P. (eds.) WAW 2014. LNCS, vol. 8882, pp. 13–22. Springer, Heidelberg (2014)
6. Bradonjić, M., Hagberg, A., Percus, A.: The structure of geographical threshold graphs. Internet Math. **5**, 113–140 (2008)
7. Cooper, C., Frieze, A.: The cover time of the preferential attachment graph. J. Comb. Theory, Ser. B **97**, 269–290 (2007)
8. Cooper, C., Frieze, A., Prałat, P.: Some typical properties of the spatial preferred attachment model. Internet Math. **10**(1–2), 116–136 (2014)
9. Demers, A., Greene, D., Hauser, C., Irish, W., Larson, J., Shenker, S., Sturgis, H., Swinehart, D., Terry, D.: Epidemic algorithms for replicated database maintenance. In: Proceedings of PODC 1987, pp. 1–12. ACM, New York (1987)
10. Doerr, B., Fouz, M., Friedrich, T.: Social networks spread rumors in sublogarithmic time. In: STOC 2011–Proceedings of the 43rd ACM Symposium on Theory of Computing, pp. 21–30. ACM, New York (2011)
11. Feige, U., Peleg, D., Raghavan, P., Upfal, E.: Randomized broadcast in networks. Random Struct. Algorithms **1**(4), 447–460 (1990)
12. Flaxman, A., Frieze, A., Vera, J.: A geometric preferential attachment model of networks II. Internet Math. **4**(1), 87–111 (2007)
13. Fountoulakis, N., Panagiotou, K., Sauerwald, T.: Ultra-fast rumor spreading in social networks. In: Proceedings of the Twenty-Third Annual ACM-SIAM Symposium on Discrete Algorithms, pp. 1642–1660. ACM, New York (2012)
14. Frieze, A.M., Grimmett, G.R.: The shortest-path problem for graphs with random arc-lengths. Discrete Appl. Math. **10**(1), 57–77 (1985)
15. Ganesan, G.: Size of the giant component in a random geometric graph. Annales de lInstitut Henri Poincaré **49**, 1130–1140 (2013)
16. Jacob, E., Mörters, P.: Robustness of scale-free spatial networks (2015). arXiv:1504.00618
17. Jacob, E., Mörters, P.: Spatial preferential attachment networks: power laws and clustering coefficients. Ann. Appl. Prob. **25**(2), 632–662 (2015)
18. Janssen, J., Prałat, P., Wilson, R.: Geometric graph properties of the spatial preferred attachment model. Adv. Appl. Math. **50**(2), 243–267 (2013)
19. Janssen, J., Prałat, P., Wilson, R.: Non-uniform distribution of points in the spatial preferential attachment model (2015). To appear in Internet Mathematics. arXiv:1506.06053
20. Karp, R., Schindelhauer, C., Shenker, S., Vöcking, B.: Randomized rumor spreading. In: 41st Annual Symposium on Foundations of Computer Science (FOCS 2000), pp. 565–574. IEEE Computer Society Press, Los Alamitos (2000)
21. Kempe, D., Kleinberg, J., Tardos, E.: Maximizing the spread of influence through a social network. In: Proceedings of the Ninth ACM SIGKDD International Conference on Knowledge Discovery and Data Mining, KDD 2003, pp. 137–146 (2003)
22. Leskovec, J., Kleinberg, J., Faloutsos, C.: Graph evolution: densification and shrinking diameters. ACM Trans. Knowl. Discov. Data (TKDD) **1**(1), Article 2, March 2007

23. Pastor-Satorras, R., Vespignani, A.: Epidemic dynamics and endemic states in complex networks. Phys. Rev. E **63**, 066117 (2001)
24. Penrose, M.: Random Geometric Graphs. Oxford Studies in Probability. Oxford University Press, Oxford (2003)
25. Zuev, K., Boguñá, M., Bianconi, G., Krioukov, D.: Emergence of soft communities from geometric preferential attachment. Nat. Sci. Rep. **5**, 9421 (2015)

A Note on Modeling Retweet Cascades on Twitter

Ashish Goel[1], Kamesh Munagala[2], Aneesh Sharma[3], and Hongyang Zhang[4(✉)]

[1] Department of Management Science and Engineering,
Stanford University, Stanford, USA
ashishg@stanford.edu

[2] Department of Computer Science, Duke University, Durham, USA
kamesh@cs.duke.edu

[3] Twitter, Inc., San Francisco, USA
aneesh@twitter.com

[4] Department of Computer Science, Stanford University, Stanford, USA
hongyz@stanford.edu

Abstract. Information cascades on social networks, such as retweet cascades on Twitter, have been often viewed as an epidemiological process, with the associated notion of *virality* to capture popular cascades that spread across the network. The notion of structural virality (or average path length) has been posited as a measure of global spread.

In this paper, we argue that this simple epidemiological view, though analytically compelling, is not the entire story. We first show empirically that the classical SIR diffusion process on the Twitter graph, even with the best possible distribution of infectiousness parameter, cannot explain the nature of observed retweet cascades on Twitter. More specifically, rather than spreading further from the source as the SIR model would predict, many cascades that have several retweets from direct followers, die out quickly beyond that.

We show that our empirical observations can be reconciled if we take *interests* of users and tweets into account. In particular, we consider a model where users have multi-dimensional interests, and connect to other users based on similarity in interests. Tweets are correspondingly labeled with interests, and propagate only in the subgraph of interested users via the SIR process. In this model, interests can be either *narrow* or *broad*, with the narrowest interest corresponding to a star graph on the interested users, with the root being the source of the tweet, and the broadest interest spanning the whole graph. We show that if tweets are generated using such a mix of interests, coupled with a varying infectiousness parameter, then we can qualitatively explain our observation that cascades die out much more quickly than is predicted by the SIR model. In the same breath, this model also explains how cascades can have large size, but low "structural virality" or average path length.

H. Zhang—This work was partly done when the author was an intern at Twitter, Inc.

D.F. Gleich et al. (Eds.): WAW 2015, LNCS 9479, pp. 119–131, 2015.
DOI: 10.1007/978-3-319-26784-5_10

1 Introduction

Information cascades are among the most widely studied phenomena in social networks. There is a vast literature on modeling the spread of these cascades as diffusion processes, studying the kinds of diffusion trees that arise, as well as trying to predict the global spread (or *virality*) of these cascades [4,8,9,11,12,16]. A specific example of such a diffusion process, which is the focus of this paper, are retweet cascades on Twitter.

Extant models of information cascades build on classical epidemiological models for spread of infectious diseases [5]. The simplest of these is the SIR model, where a node in the network can be in one of three states at any time: *Susceptible* (S); *Infected* (I); and *Recovered* (R). Nodes in the network switch their states due to infections transmitted over the network, and the rate of these infections is governed by an infectiousness parameter, p. The SIR model unfolds via the following process: all nodes are initially in state S except the source (or a set of nodes called the "seed set"), which is in state I. Every node which is in state I infects each of its neighbors independently with probability p, before moving itself to state R. If a node in state S gets infected, it moves to state I. This process naturally quiesces with all nodes settling in their final state, and all nodes that were ever in state I are considered to have acquired the infection. There is a natural and trivial mapping of this model to information cascades, where the infectiousness parameter p serves to measure the *interestingness* of the piece of information, in our case, a tweet. In epidemiology, the goal is to differentiate infections that die out quickly from those that spread to the whole network; analogously, information cascades are deemed *viral* if their global reach is large.

The above view of information cascades as the spreading of content through the network is intuitively and analytically appealing. In fact, Goel *et al.* show that when simulated on a scale-free graph, the SIR model statistically mimics important properties of retweet cascades on Twitter. In particular, they use *structural virality*, or average path length in the diffusion tree, as a quantitative measure of "infectiousness" of a cascade, and show that the distribution of cascade sizes (number of users that retweet a tweet plus the author of the tweet) and structural virality are statistically similar to that from the simulations. On the other hand, these empirical studies also show that cascades observed in Twitter are mostly shallow and exceedingly rare: Goel *et al.* [7] show there are no viral cascades in a corpus of a million tweets; and in subsequent work [6], show that viral cascades do indeed exist if the corpus size is increased to a billion tweets. This data contrasts with the observation that social networks like Twitter have a power-law degree distribution [13], and these networks should have low epidemic threshold, so that even with low infectiousness parameter p, most cascades should be viral [1,2]. Therefore, explaining the low frequency of viral events on Twitter via an SIR model requires that the infectiousness parameter be quite low almost all the time. Finally, this result also begs the question of whether modeling viral events if even of any interest if these events are so rare.

We therefore ask: *Is there something fundamental about real-world information cascades, particularly those on Twitter, that is not captured by the simple SIR model?* Though this question is about a specific social network, and a specific (simplistic) epidemiological model, even understanding this via suitably designed experiments is challenging, and has not been performed before.

1.1 Our Contributions

In the process of answering the above question, we make the following contributions.

Evaluating Epidemic Models Through Twitter Network. Our main contribution is to show that the SIR model is a *poor* fit for information flow on Twitter. We show this by empirically testing the hypothesis that retweet cascades on Twitter propagate using the SIR process. Our null hypothesis is that each cascade has an underlying infectiousness p (that could be different for different cascades), and conditioned on receiving the tweet, a user retweets it with probability p. We compare the value of p that we obtain by best-fit for the users directly connected to the source of the tweet (level 1 followers), and those who receive the tweet from a direct follower of the source (level 2 followers). Using a corpus of 8 million cascades, we develop a statistical test to show that these two values of p are different – the second level value is significantly smaller than the first. The technically interesting part of this analysis is the fact that most cascades are shallow. Thus, many tweets generate very few retweets at the first level, and this number dictates the number of tweet impressions and retweets at the second level. The SIR model therefore corresponds to a stochastic process for the retweets that has very low mean but potentially very high variance because of the skewed degree distribution of the graph. We have to therefore devise a statistical test that works around this high variance. Apart from this statistical test, at a coarse level, we find that the median value of first level infection probability is 0.00046, while the median value of second level infection probability is 0 (in other words, half of the tweets do not have second level retweets!). Even among the tweets that have at least 1000 impressions at the first level, more than 80 % of them, have that first level p is at least twice the second level p. This suggests that, rather than spreading further from the source, a cascade typically dies out quickly within a few hops.[1] This echoes with the observation that most of the cascades tend to be star-like trees [16]. It also suggests an explanation for truly viral cascades being so rare [6].

Interest-based SIR Model. Since the SIR model assumption of fixed propagation probability per cascade is statistically violated on Twitter, we propose an alternative model for retweet cascades. In particular, we present a tweet propagation model that takes *interests* of users and tweets into account. In order to do

---

[1] Indeed, the median of first level impressions is 175, while the median of second level impressions is 29!

this, we revisit a Kronecker graph-based model for social networks first considered in [3]. In this attribute based model, users have attribute vectors in some d-dimensions, and interests are specified by a subset of these dimensions along with their attribute values. If fewer dimensions are specified, these interests are *broad* and encompass many users; if many dimensions are specified, these interests are *narrow* with a shallow component around the source. Tweets are also correspondingly labeled with interests, and propagate only in the *subgraph of interested users* via a SIR process with infectiousness drawn from a distribution. We show that if tweets are generated using such a mix of narrow and broad interests, then this coupled with a varying infectiousness parameter can qualitatively explain the level-one infectiousness being larger than subsequent levels. As a simple intuition, observe that cascades corresponding to narrow interests only reside in their shallow subgraphs, while those corresponding to broad interests can be "viral" in the usual sense.

As mentioned above, Goel *et al.* [6] define the notion of *structural virality*, or average path length of a cascade as a measure of its virality. They show that this measure is uncorrelated with the size of the cascade, except when structural virality is large. The proposed explanation in their work is an SIR model on a scale-free graph with extremely low infectiousness parameter. Our model leads to a different explanation: cascades corresponding to narrow interests have low structural virality, but can have large size. This explanation does not depend on any specific setting of the infectiousness parameter, and is therefore of independent interest. Finally, we show that cascades arising for broad interests can have large structural virality, but our model would predict a large expected size as well, which again matches previous empirical findings.

1.2   Related Work

Epidemic models on social networks have received a lot of attention in the past decade, and we won't attempt to review the large literature here. Instead, we point the reader to a small set of representative papers and the excellent survey articles and books on the topic [4,5,9,11,12,14]. Despite all the attention on studying diffusion, there has been relatively little work evaluating epidemic models on social networks such as Twitter [6,14,19]. In particular, we believe that the empirical testing of structural properties of cascades on the Twitter graph (as opposed to a specific generative model) is unique to our work.

Part of the reason, as has been pointed out in [6], is that only recently have large datasets of information contents become available. In the same work, the authors defined the notion of structural virality and observe that it is very rare to observe structurally viral cascades, but they can find these rare cascades by obtaining a large collection of tweets. By carefully choosing the infectiousness parameter of the SIR model on a power law network, they are able to reproduce many empirical statistics of the observed cascades distribution, such as the probability that a piece of content gains at least 100 adopters, and the mean structural virality. However, they also point out that other important statistics does not match with the empirical distribution. For example, the variance is much smaller in the simulated model, compared to the empirical distribution. We present an alternative interest-based model for explaining the same phenomena, while comprehensively refuting the SIR hypothesis.

Similarly, Leskovec et al. [16] were able to fit cascade sizes and degree distributions of a large collection of blogs, with the SIS model defined by an infectiousness parameter. We also want to mention a study of user adoption on Facebook, Ugander *et al.* [19] find that the probability of users joining Facebook is dependent on the number of connected components in an user's ego network (or neighborhood graph), rather than by the size of the ego network. Note that this work studied user adoption rather than content diffusion, but the observation that sub-structures in the network can dominate network size for adoption is in general agreement with our proposed model.

2 Evaluating the SIR Model on the Twitter Network

In this section, we describe our evaluation of the simple SIR model on eight million retweet cascades observed on Twitter. These retweet cascades are collected from a single week and each cascade is restricted to be started by a user based in the US. In our analysis, we have excluded tweets posted by Twitter accounts that are likely to be spammers using an internal quality detection tool.[2] For each tweet, we collect the information described in Table 1. Note that we use the number of followers of a user as a proxy for the number of impressions of the user's tweets. While we could also count impressions directly on Twitter, this would not correctly represent the significant fraction of users that visit Twitter through third-party clients. All the information described in Table 1 could be collected through the public Twitter APIs.[3] While we used Twitter's internal spam detection mechanism to filter away potential spam users, we believe that exploiting well-known features (for example pagerank values) would also achieve the same results for our task.

2.1 Defining the Null Hypothesis

Let us fix a given set of tweets T. For each tweet $t \in T$, let $p_1(t)$ and $p_2(t)$ denote the underlying retweet rate at the first level and second level of the Twitter graph, respectively. Note that these parameters are fixed but unknown for any given tweet. The dependence of p_1 and p_2 on t models the fact that different tweets can have different infectiousness. Our null hypothesis is that $p_1(t) = p_2(t)$ for all $t \in T$, which corresponds to cascade propagation via the simple SIR model. A different, but equivalent view of the null hypothesis is that it posits $p_1(t)$ is drawn from some distribution, and conditioned on this, we set $p_2(t) = p_1(t)$.

---

[2] A lot of spam tweets have star-like cascade structure that may significantly impact the experiment results while not representing general user behavior.

[3] https://dev.twitter.com/streaming/public.

Table 1. A list of observed information for a tweet τ, posted by a node s. Let $N_1(\tau)$ denote the set of nodes that follow the node s. Let $R_1(\tau)$ denote the subset of nodes among $N_1(\tau)$ that retweet the tweet τ. And let $N_2(\tau)$ denote the set of nodes that follow any nodes in $R_1(\tau)$.

v_1	Number of followers of the source node (the size of $N_1(\tau)$)
r_1	Number of retweets among the set of nodes $N_1(\tau)$ (the size of $R_1(\tau)$)
v_2	Number of nodes that follow any nodes in $R_1(\tau)$ (the size of $N_2(\tau)$)
r_2	Number of retweets among the set of nodes in $N_2(\tau)$

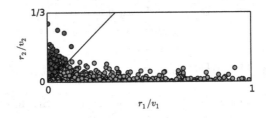

Fig. 1. A scatter plot of ten thousand sampled tweets. The y-axis has been truncated since there are no points beyond $1/3$ in the samples.

The stochastic process, given a tweet t and corresponding underlying $p_1(t)$ and $p_2(t)$ unfolds as follows (we omit t for notational convenience): let the value v_1 be a (non-random) parameter associated with the tweet source. Then $r_1 \sim B(v_1, p_1)$ is a Binomial random variable with parameters v_1 and the unknown p_1. We will *assume* that v_2 (th) is nonzero whenever r_1 is nonzero. Since v_2 is defined as the total number of followers among those who retweet the source tweet, if this value is zero with r_1 being non-zero, then the source user is very likely to be a spammer. However, since we eliminated spam sources in our filtering step, this event is very unlikely in our dataset. Now, r_2 is a random variable that is generated according to $B(v_2, p_2)$. Note that we are modelling r_2 as a Binomial random variable, since it is easier to present than the SIR process. As a matter of fact, there is no difference to our conclusions if r_2 is generated according to the SIR process. The reason for that is Lemma 1 will continue to hold under the SIR process. We observe a realization of the random variables, $v_2, r_1,$ and r_2.

2.2   Refuting the SIR Model

We will now refute the null hypothesis, *i.e.*, show that $p_1(t) > p_2(t)$ for almost all $t \in T$. Observe that if $r_1(t)$ and $r_2(t)$ are sufficiently large, then by standard concentration bounds, $\frac{r_1(t)}{v_1(t)}$ will be a good approximation to $p_1(t)$, and likewise for $p_2(t)$. A natural approach is therefore to compare the empirical average of $\frac{r_1(t)}{v_1(t)}$ over $t \in T$ to the empirical average of $\frac{r_2(t)}{v_2(t)}$. If these are different, that would refute $p_1(t) = p_2(t)$ for all $t \in T$. In Fig. 1, we plot these empirical values, and this provides some evidence that the null hypothesis is false. However, this approach is not quite statistically rigorous.

Specifically, the problem with this approach is that when $r_1(t)$ is zero, then $v_2(t)$ is zero and $p_2(t)$ remains undefined. However, if we filter away any tweet whose $r_1(t) = 0$, then we could potentially bias the estimation of $p_1(t)$ as well. To overcome this issue, we will correct the bias by subtracting a corresponding factor in $\frac{r_1(t)}{v_1(t)}$.

In the lemmas and definitions below, the expectation is over the stochastic process described above, where v_2, r_1, r_2 are random variables. For each tweet $t \in T$ we define the following random variables:

$$X_2(t) = \begin{cases} r_2(t)/v_2(t) \text{ if } v_2(t) > 0 \\ 0 \text{ if } v_2(t) = 0 \end{cases} \tag{1}$$

$$X_1(t) = r_1(t)/v_1(t) - f_0(t) \tag{2}$$

where $f_0(t) = (\frac{v_1(t)}{v_1(t)+1})^{v_1(t)+1}/v_1(t)$.

Lemma 1. *Under the null hypothesis that $p_1(t) = p_2(t)$, we have $\mathbb{E} X_2(t) \geq \mathbb{E} X_1(t)$, for any $t \in T$.*

Proof. Note that

$$\mathbb{E} X_2(t) = p(t) \Pr(v_2(t) \neq 0) = p(t) \Pr(r_1(t) \neq 0),$$

by our assumption that $v_2(t) = 0$ if and only if $r_1(t) = 0$. Further,

$$\mathbb{E} X_1(t) = p(t) - f_0(t)$$

The conclusion follows since:

$$p(t) \Pr(r_1(t) = 0) = p(t) \times (1 - p(t))^{v_1(t)} \leq f_0(t).$$

where the last inequality is obtained by observing the maximum value of the function $p(t) \times (1 - p(t))^{v_1(t)}$ of $p(t)$.

For any subset T of tweets, let $\chi_1 = \sum_{t \in T} X_1(t)$ and $\chi_2 = \sum_{t \in T} X_2(t)$. We compute the observed values of χ_1 and χ_2 for several different buckets of tweets T, grouped by ranges over number of first level impressions. These buckets are shown in Table 2. Based on the second and third columns, we conclude that the average observed X_2 is less than the average observed X_1, thereby contradicting the null hypothesis.

Now we examine the significance of the above finding. The idea is that since both χ_1 and χ_2 are sums of independent random variables in the range $[0, 1]$, the observed values should be concentrated around the mean value. While we don't know the mean values, $\mathbb{E} \chi_1$ and $\mathbb{E} \chi_2$, we can obtain an upper bound of the desired probability, by maximizing over all possible values of $\mathbb{E} \chi_1$ and $\mathbb{E} \chi_2$,

Table 2. Experimental results for several different buckets of tweets. See main text for more details.

v_1	Number of tweets	χ_1	χ_2	p-value
$(0, \infty)$	3766 k	3017	836	0.0
$(100, 1000)$	359 k	690	109	10^{-100}
$(1000, 10000)$	2133 k	1830	531	10^{-150}
$(10000, \infty)$	1274 k	477	195	10^{-30}

subject to the null hypothesis, Lemma 1. This is summarized in the following Lemma:

Lemma 2. *For a set of tweets T with observed values of $\chi_1 \geq \chi_2$, the probability that such an observation could happen under the null hypothesis, $p_1(t) = p_2(t)$ for all $t \in T$, can be upper bounded by:*

$$2 \exp(-\frac{2\sqrt{2(\chi_1^2 + \chi_2^2)} - 2\chi_1 - 2\chi_2}{3}).$$

Proof. Let $t_1 = \mathbb{E}\,\chi_1$ and $t_2 = \mathbb{E}\,\chi_2$. By Chernoff bound (cf Corollary 4.6 [18]),

$$\Pr(|\chi_1 - t_1| \geq \delta_1 t_1) \leq 2\exp(-t_1\delta_1^2)/3$$
$$\Pr(|\chi_2 - t_2| \geq \delta_2 t_2) \leq 2\exp(-t_2\delta_2^2)/3$$

Hence

$$\max_{t_2 \geq t_1 > 0} \Pr(|\chi_1 - t_1| \geq \delta_1 t_1, |\chi_2 - t_2| \geq \delta_2 t_2)$$
$$\leq \max_{t_2 \geq t_1 > 0} 2\exp(-(t_1\delta_1^2 + t_2\delta_2^2)/3)$$
$$= \max_{t_2 \geq t_1 > 0} 2\exp(-(\frac{\chi_1^2}{t_1} + t_1 + \frac{\chi_2^2}{t_2} + t_2 - 2\chi_1 - 2\chi_2)/3) \qquad (3)$$

Consider two cases,

1. if $t_2 \leq \chi_1$, then we know that $\frac{\chi_1^2}{t_1} + t_1 \geq \frac{\chi_1^2}{t_2} + t_2$, and (3) can be upper bounded by

$$2\exp(-\frac{2\sqrt{2(\chi_1^2 + \chi_2^2)} - 2\chi_1 - 2\chi_2}{3})$$

when $t_2 = t_1 = \sqrt{\frac{\chi_1^2 + \chi_2^2}{2}}$.

2. if $t_2 > \chi_1$, then we know that $\frac{\chi_1^2}{t_1} + t_1 \geq 2\chi_1$, and $\frac{\chi_1^2}{t_1} + t_1 \geq \frac{\chi_1^2}{\chi_1} + \chi_1$. Then (3) can be upper bounded by

$$2\exp(-(\frac{\chi_1^2}{\chi_1} + \chi_1 - 2\chi_2)/3)$$

when $t_1 = t_2 = \chi_1$. And it's not hard to check that this is smaller than the bound obtained in the first case.

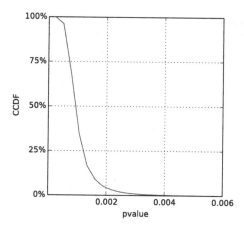

Fig. 2. The histogram of p-values obtained from 10000 random subset of tweets. Each random subset contains 1 % of all tweets.

We compute these probabilities and show them as p-values in Table 2. This shows that the observed χ_1, χ_2 are highly unlikely under the null hypothesis.

Finally, we note that the above analysis does not necessarily show that $p_1(t) > p_2(t)$ for almost all t in our corpus. To address this concern, we randomly sample 1 % of the tweets, run the same analysis, and repeat for 10000 times. Figure 2 plots the histogram of p-values that we obtain. Since we observed consistently low p-values among all the samples, this shows that the null hypothesis of $p_1(t) = p_2(t)$ for all $t \in T$ is very unlikely to hold in our dataset. In fact, our analysis shows that $p_1(t)$ is almost always bigger than $p_2(t)$.

3   An Interest Based Model for Tweet Propagation

We now show that the above empirical observations are consistent with the following model: Users have interests and connect to other users based on similarities in interests. Each tweet corresponds to an interest (either a broad interest or a narrow interest) and is retweeted only by users with the corresponding interest. We formalize this model below, and show how it can qualitatively explain our observations.

We adapt the Kronecker interest model formulated in [3]. This is based on the Kronecker social graph, which has been studied as a reasonable theoretical model for social networks [15,17]. We note that some of the model assumptions below are not an exact fit for social networks; nevertheless, this model captures most high-level statistical properties observed in reality, in addition to being easy to interpret. In our model, parametrized by a small number K, there are $|V| = n$ users, and $d = \log_K n$ attributes, each with K possible values from the set $S = \{a_1, a_2, \ldots, a_K\}$. Each node $u \in V$ maps to a d-dimensional vector of attribute values $(u_1, u_2, \ldots, u_d)$, where each $u_i \in S$. Therefore, $|V| = K^d = n$.

Treat the values in S as the K vertices of an undirected *seed graph* G_0, and denote the adjacency matrix of this graph as A. Assume $A[a_s, a_s] = 1$ for $1 \leq s \leq K$.

For each $u = (u_1, u_2, \ldots, u_d)$ and $v = (v_1, v_2, \ldots, v_d)$, the edge (u, v) exists iff $A[u_j, v_j] = 1$ for all $j = 1, 2, \ldots, d$. We define an interest as a set of pairs of attribute dimensions and their values, where a generic interest $i \in I$ has the following form:

$$i = \{\langle j_1, a_{j_1} \rangle, \langle j_2, a_{j_2} \rangle, \ldots, \langle j_r, a_{j_r} \rangle\} \quad \text{where} \quad j_1, j_2, \ldots, j_r \leq K \text{ and } r \leq d$$

The *consumers* of this interest are defined as:

$$C_i = \{u = (u_1, u_2, \ldots, u_d) \mid A[u_j, a_j] = 1 \ \forall \langle j, a_j \rangle \in i\}$$

Similarly, the producers of this interest are defined as:

$$P_i = \{u = (u_1, u_2, \ldots, u_d) \mid u_j = a_j \ \forall \langle j, a_j \rangle \in i\}$$

The above interest model has the following interpretation. Since each interest is specified by a subset of attributes along with their values, the graph G_0 and adjacency matrix A specify which interests are related, i.e. which interests specify an *interested in* relationship. We classify interests are *narrow* or *broad*. The narrowest interests have $|i| = d$, and the broadest interest has $|i| = 0$. Further, these interests have a natural hierarchical structure, where the *broader* interests are those specified by fewer attributes. Also note that a producer of an interest needs to align with it's attribute values on all the relevant attribute dimensions, while a consumer of an interest only needs to be *interested in* those attribute values in the relevant attribute dimensions.

We parametrize the tweet propagation process by two distributions: There is an interest distribution F and a SIR parameter distribution G. We choose an interest i at random from distribution F; choose a producer u uniformly at random from P_i, and choose an infectiousness p at random from G. The tweet originates at u, and propagates using the SIR model with parameter p on the subgraph induced by C_i.

We now perform some calculations to understand the behavior of this process for various interest sizes. In order to simplify these calculations, we assume G_0 is regular with degree w, and denote $A = w^d$ as the degree of each user. We assume $A \gg w$. Note that G_0 has K vertices, so $w \leq K$. We denote $d - |i| = s$ as the *size* of the interest. We further assume that the infectiousness parameter p is small so that $wp \ll 1$; on the other hand, we assume it is large enough that $Ap \gg 1$. We note that these assumptions are only to derive simple formulas that can be qualitatively interpreted. We need to use more nuanced parameter settings to model real social networks, but these will not affect the high-level qualitative nature of our conclusions.

Narrowest Interests, $s = 0$ In this case, $|P_i| = 1$, so that there is one user u who is a potential producer. This user is directly connected to all users in C_i. Therefore, for any p, the size of the cascade is Ap, and the structural virality is exactly 2.

Narrow Interest, $s = 1$ In this case, $|P_i| = K$, and these producers are connected as G_0. Assume all these producers have the first $d - 1$ coordinates of their attribute vector fixed to one value, and the final coordinate taking one of K possible values. The consumers C_i are all the neighbors of P_i. For small enough p, let $wp = \delta \in (0, 1)$. Then we approximately have Size $= \frac{A}{w}\delta(1 + \delta)$, and $SV = 2 + \frac{\delta}{2}$. In this case, though structural virality grows very slowly with size, a large structural virality implies a large size but not necessarily the other way around.

Broad Interest, $s = d$ In this case, $|P_i| = n$. Assuming $Ap \gg 1$, the expected size of the cascade is $(Ap)^h$, where $h = \log_A n$ is the depth of the process. The structural virality is $2h$ regardless of p. Therefore, for broad interests with moderate infectiousness p, we expect a high value of structural virality, and a correspondingly high value of size. Therefore, in our model, a high value of structural virality corresponds to a broader interest, and these cascades also have large size.

4   Conclusion

In this paper, we performed an empirical examination of the SIR epidemic model on a large selection of retweet cascades on Twitter. The experimental results refute the null hypothesis, and show that the SIR model does not fit the empirical observations. This is because retweet rates decrease as a cascade propagates further from the source, contradicting the fixed probability per cascade assumption in the SIR model. We also proposed an alternative interest-based diffusion model, where users retweet based on overlapping interests with a tweet. It is an interesting future challenge to empirically test the interest-based diffusion model. Indeed, in preliminary experiments we often found that structurally viral cascades correspond to "broad" topics that also have a very large size. In particular, we tweets containing jokes, appeals for finding a lost person, and "not safe for work" (NSFW) content are common among large structurally viral retweet cascades. On the other hand, tweets that correspond to "narrow" topics (niche sports and other topical content) usually have small structural virality. We leave it to future work to validate these observations on a large scale.

We also emphasize that our work is specific to the flow on information in social networks such as Twitter, and on fitting the simple SIR model (with possibly different levels of infectiousness or interestingness for different tweets) to it. We view this work as one further step towards validating simple models for information spreading. Given the format of retweets on Twitter where multiple retweets to a user can be suppressed, we have not considered threshold models (such as in [10]) that are based on a user receiving multiple copies of the message from different sources. We note that such threshold models have been extensively investigated in other diffusion contexts such as adoption of new technologies, and are likely appropriate for spread of information cascades in other social media. This makes it a good topic for future investigation. We also note that the interest-based model, coupled with SIR on the appropriate interest subgraph, is only one possible explanation for our observations. It is an interesting research direction to see if there are other possible explanations, such as local structure in networks, epidemic thresholds, etc. that can be empirically validated. Finally, an interesting direction is to explore alternative notions of virality other than structural virality. In particular, is there a way to capture "viral" events that are specific to a group of friends, or inside a community? We believe that understanding these questions will also provide new insights for content recommendation and targeting on social networks.

Acknowledgment. We are grateful to the anonymous reviewers for very helpful feedbacks. Goel and Zhang are supported by DARPA GRAPHS program via grant FA9550-12-1-0411. Munagala is supported in part by NSF grants CCF-1348696, CCF-1408784, and IIS-1447554, and by grant W911NF-14-1-0366 from the Army Research Office (ARO).

References

1. Berger, N., Borgs, C., Chayes, J.T., Saberi, A.: On the spread of viruses on the internet. In: Proceedings of the ACM-SIAM Symposium on Discrete Algorithms (SODA), pp. 301–310. Society for Industrial and Applied Mathematics (2005)
2. Boguná, M., Pastor-Satorras, R., Vespignani, A.: Absence of epidemic threshold in scale-free networks with degree correlations. Phys. Rev. Lett. **90**(2), 028701 (2003)
3. Bosagh Zadeh, R., Goel, A., Munagala, K., Sharma, A.: On the precision of social and information networks. In: Proceedings of the ACM Conference on Online Social Networks (COSN), pp. 63–74 (2013)
4. Cheng, J., Adamic, L., Dow, P.A., Kleinberg, J.M., Leskovec, J.: Can cascades be predicted? In: Proceedings of the 23rd World Wide Web Conference (WWW), pp. 925–936 (2014)
5. Easley, D., Kleinberg, J.: Networks, Crowds, and Markets: Reasoning About a Highly Connected World. Cambridge University Press, New York (2010)
6. Goel, S., Anderson, A., Hofman, J., Watts, D.: The structural virality of online diffusion. Management Science (2015)
7. Goel, S., Watts, D.J., Goldstein, D.G.: The structure of online diffusion networks. In: Proceedings of the ACM EC, pp. 623–638 (2012)
8. Golub, B., Jackson, M.O.: How homophily affects diffusion and learning in networks. The Quarterly Journal of Economics (2012)
9. Gomez-Rodriguez, M., Leskovec, J., Krause, A.: Inferring networks of diffusion and influence. In: Proceedings of the SIGKDD Conference on Knowledge Discovery and Data Mining (KDD), pp. 1019–1028 (2010)
10. Kempe, D., Kleinberg, J., Tardos, É.: Maximizing the spread of influence through a social network. In: Proceedings of the SIGKDD Conference on Knowledge Discovery and Data Mining (KDD), pp. 137–146 (2003)
11. Kempe, D., Kleinberg, J.M., Tardos, É.: Influential nodes in a diffusion model for social networks. In: Caires, L., Italiano, G.F., Monteiro, L., Palamidessi, C., Yung, M. (eds.) ICALP 2005. LNCS, vol. 3580, pp. 1127–1138. Springer, Heidelberg (2005)
12. Kleinberg, J.: Cascading behavior in networks: algorithmic and economic issues. In: Nisan, N., Roughgarden, T., Tardos, E., Vazirani, V. (eds.) Algorithmic Game Theory, pp. 613–632. Cambridge University Press, UK (2007)

13. Kwak, H., Lee, C., Park, H., Moon, S.: What is twitter, a social network or a news media? In: Proceedings of the 19th International Conference on World Wide Web, pp. 591–600. ACM (2010)

14. Leskovec, J., Adamic, L.A., Huberman, B.A.: The dynamics of viral marketing. ACM Trans. Web (TWEB) 1(1), 5 (2007)

15. Leskovec, J., Chakrabarti, D., Kleinberg, J., Faloutsos, C., Ghahramani, Z.: Kronecker graphs: an approach to modeling networks. J. Mach. Learn. Res. 11, 985–1042 (2010)

16. Leskovec, J., McGlohon, M., Faloutsos, C., Glance, N.S., Hurst, M.: Patterns of cascading behavior in large blog graphs. In: Symposium on Data Mining (SDM), vol. 7, pp. 551–556 (2007)

17. Mahdian, M., Xu, Y.: Stochastic Kronecker graphs. Random Struct. Algorithms 38(4), 453–466 (2011)

18. Mitzenmacher, M., Upfal, E.: Probability and Computing: Randomized Algorithms and Probabilistic Analysis. Cambridge University Press, New York (2005)

19. Ugander, J., Backstrom, L., Marlow, C., Kleinberg, J.: Structural diversity in social contagion. Proc. Natl. Acad. Sci. (PNAS) 109(16), 5962–5966 (2012)

The Robot Crawler Number of a Graph

Anthony Bonato[1]([✉]), Rita M. del Río-Chanona[3], Calum MacRury[2],
Jake Nicolaidis[1], Xavier Pérez-Giménez[1], Paweł Prałat[1], and Kirill Ternovsky[1]

[1] Ryerson University, Toronto, Canada
abonato@ryerson.ca
[2] Dalhousie University, Halifax, Canada
[3] Universidad Nacional Autónoma de Mexico, Mexico City, Mexico

Abstract. Information gathering by crawlers on the web is of practical interest. We consider a simplified model for crawling complex networks such as the web graph, which is a variation of the robot vacuum edge-cleaning process of Messinger and Nowakowski. In our model, a crawler visits nodes via a deterministic walk determined by their weightings which change during the process deterministically. The minimum, maximum, and average time for the robot crawler to visit all the nodes of a graph is considered on various graph classes such as trees, multipartite graphs, binomial random graphs, and graphs generated by the preferential attachment model.

1 Introduction

A central paradigm in web search is the notion of a *crawler*, which is a software application designed to gather information from web pages. Crawlers perform a walk on the web graph, visiting web pages and then traversing links as they explore the network. Information gathered by crawlers is then stored and indexed, as part of the anatomy of a search engine such as Google or Bing. See [10,16,25] and the book [22] for a discussion of crawlers and search engines.

Walks in graph theory have been long-studied, stretching back to Euler's study of the Königsberg bridges problem in 1736, and including the travelling salesperson problem [3] and the sizeable literature on Hamiltonicity problems (see, for example, [28]). An intriguing generalization of Eulerian walks was introduced by Messinger and Nowakowski in [23], as a variant of graph cleaning processes (see, for example, [2,24]). The reader is directed to [8] for an overview of graph cleaning and searching. In the model of [23], called the *robot vacuum*, it is envisioned that a building with dirty corridors (for example, pipes containing algae) is cleaned by an autonomous robot. The robot cleans these corridors in a greedy fashion, so that the next corridor cleaned is always the "dirtiest" to which it is adjacent. This is modelled as a walk in a graph. The robot's initial position is any given node, with the initial weights for the edges of the graph G being $-1, -2, \ldots, -|E(G)|$ (each edge has a different value). At every step of

Research supported by grants from NSERC, MITACS Inc. and Ryerson University.

the walk, the edges of the graph will be assigned different weights indicating the last time each one was cleaned (and thus, its level of dirtiness). It is assumed that each edge takes the same length of time to clean, and so weights are taken as integers. In such a model, it is an exercise to show that for a connected graph, one robot will eventually clean the graph (see [23]).

Let $s(G)$ and $S(G)$ denote the minimum and maximum number of time-steps over all edge weightings, respectively, when every edge of a graph G has been cleaned. As observed in [23], if G is an Eulerian graph, then we have that $s(G) = |E(G)|$, and moreover the final location of the robot after the first time every edge has been cleaned is the same as the initial position. Li and Vetta [20] gave an interesting example where the robot vacuum takes exponential time to clean the graph. Let S_e be the maximum value of S over all connected graphs containing exactly e edges. It is proven in [20] that there exists an explicit constant $d > 0$ such that, for all e, $S_e \geq d(3/2)^{e/5} - 1/2$. Moreover, $S_e \leq 3^{e/3+1} - 3$. An analogous result was independently proven by Copper et al. [13] who analyzed a similar model to the robot vacuum process. The "self-stabilization" found in robot vacuum is also a feature of so-called *ant algorithms* (such as the well-known *Langton's ant* which is capable of simulating a *universal Turing machine*; see [15]). The robot vacuum model can be regarded as an undirected version of the *rotor-router* model; see [27,29].

In the present work, we provide a simplified model of a robot crawler on the web, based on the robot vacuum paradigm of [20,23]. In our model, the crawler cleans nodes rather than edges. Nodes are initially assigned unique non-positive integer weights from $\{0, -1, -2, \ldots, -|V(G)| + 1\}$. In the context of the web or other complex networks, weights may be correlated with some popularity measure such as in-degree or PageRank. The robot crawler starts at the dirtiest node (that is, the one with the smallest weight), which immediately gets its weight updated to 1. Then at each subsequent time-step it moves greedily to the dirtiest neighbour of the current node. On moving to such a node, we update the weight to the positive integer equalling the time-step of the process. The process stops when all weights are positive (that is, when all nodes have been cleaned). Note that while such a walk by the crawler may indeed be a Hamilton path, it usually is not, and some weightings of nodes will result in many re-visits to a given node. Similar models to the robot crawler have been studied in other contexts; see [18,21,27].

A rigorous definition of the robot crawler is given in Sect. 2. We consider there the minimum, maximum, and average number of time-steps required for the robot crawler process. We give asymptotic (and in some cases exact) values for these parameters for paths, trees, and complete multi-partite graphs. In Sect. 3, we consider the average number of time-steps required for the robot crawler to explore binomial random graphs. The robot crawler is studied on the preferential attachment model, one of the first stochastic models for complex networks, in Sect. 4. We conclude with a summary and a list of open problems for further study. Due to lack of space, some of the proofs are omitted from this extended abstract and deferred to the extended version.

Throughout, we consider only finite, simple, and undirected graphs. For a given graph $G = (V, E)$ and $v \in V$, $N(v)$ denotes the neighbourhood of v and $\deg(v) = |N(v)|$ its degree. For background on graph theory, the reader is directed to [28]. For a given $n \in \mathbb{N}$, we use the notation $B_n = \{-n + 1, -n + 2, \ldots, -1, 0\}$ and $[n] = \{1, 2, \ldots, n\}$. All logarithms in this paper are with respect to base e. We say that an event A_n holds *asymptotically almost surely* (*a.a.s.*) if it holds with probability tending to 1 as n tends to infinity.

2   The Robot Crawler Process: Definition and Properties

We now formally define the robot crawler process and the various robot crawler numbers of a graph. Some proofs are omitted owing to space constraints, and will appear in the full version of the paper. The *robot crawler* $\mathcal{RC}(G, \omega_0) = \left((\omega_t, v_t)\right)_{t=1}^{L}$ of a connected graph $G = (V, E)$ on n nodes with an *initial weighting* $\omega_0 : V \to B_n$, that is a bijection from the node set to B_n, is defined as follows.

1. Initially, set v_1 to be the node in V with weight $\omega_0(v_1) = -n + 1$.
2. Set $\omega_1(v_1) = 1$; the other values of ω_1 remain the same as in ω_0.
3. Set $t = 1$.
4. If all the weights are positive (that is, $\min_{v \in V} \omega_t(v) > 0$), then set $L = t$, stop the process, and return L and $\mathcal{RC}(G, \omega_0) = \left((\omega_t, v_t)\right)_{t=1}^{L}$.
5. Let v_{t+1} be the dirtiest neighbour of v_t. More precisely, let v_{t+1} be such that

$$\omega_t(v_{t+1}) = \min\{\omega_t(v) : v \in N(v_t)\}.$$

6. $\omega_{t+1}(v_{t+1}) = t + 1$; the other values of ω_{t+1} remain the same as in ω_t.
7. Increment to time $t + 1$ and return to 4.

If the process terminates, then define

$$\mathrm{rc}(G, \omega_0) = L,$$

that is $\mathrm{rc}(G, \omega_0)$ is equal to the number of steps in the *crawling sequence* $(v_1, v_2, \ldots, v_L)$ (including the initial state) taken by the robot crawler until all nodes are clean; otherwise $\mathrm{rc}(G, \omega_0) = \infty$. We emphasize that for a given ω_0, all steps of the process are deterministic. Note that at each point of the process, the weighting ω_t is an injective function. In particular, there is always a unique node v_{t+1}, neighbour of v_t of minimum weight (see step (4) of the process). Hence, in fact, once the initial configuration is fixed, the robot crawler behaves like a cellular automaton. It will be convenient to refer to a node as *dirty* if it has a non-positive weight (that is, it has not been yet visited by the robot crawler), and *clean*, otherwise.

The next observation that the process always terminates in a finite number of steps is less obvious, but we omit the proof owing to space constraints.

Theorem 1. *For a connected graph $G = (V, E)$ on n nodes and a bijection $\omega_0 : V \to B_n$, $\mathcal{RC}(G, \omega_0)$ terminates after a finite number of steps; that is, $\mathrm{rc}(G, \omega_0) < \infty$.*

The fact that every node in a graph will be eventually visited inspires the following definition. Let $G = (V, E)$ be any connected graph on n nodes. Let Ω_n be the family of all initial weightings $\omega_0 : V \to B_n$. Then

$$\mathrm{rc}(G) = \min_{\omega_0 \in \Omega_n} \mathrm{rc}(G, \omega_0) \qquad \text{and} \qquad \mathrm{RC}(G) = \max_{\omega_0 \in \Omega_n} \mathrm{rc}(G, \omega_0).$$

In other words, $\mathrm{rc}(G)$ and $\mathrm{RC}(G)$ the are minimum and maximum number of time-steps, respectively, needed to crawl G, over all choices of initial weightings. Now let $\overline{\omega}_0$ be an element taken uniformly at random from Ω_n. Then we have the average case evaluated as

$$\overline{\mathrm{rc}}(G) = \mathbb{E}\left[\mathrm{rc}(G, \overline{\omega}_0)\right] = \frac{1}{|\Omega_n|} \sum_{\omega_0 \in \Omega_n} \mathrm{rc}(G, \omega_0).$$

The following result is immediate. (Part 5. follows from the observation that, if a node v is cleaned by the robot crawler $\Delta + 1$ times within an interval of time-steps, then every neighbour of v must be cleaned at least once during that interval.)

Lemma 1. *Let G be a connected graph of order n, maximum degree Δ, and diameter d. Let C_n and K_n denote the cycle and the clique of order n, respectively.*

1. $\mathrm{rc}(G) \leq \overline{\mathrm{rc}}(G) \leq \mathrm{RC}(G)$.
2. $\mathrm{rc}(K_n) = \overline{\mathrm{rc}}(K_n) = \mathrm{RC}(K_n) = n$.
3. $\mathrm{rc}(C_n) = \overline{\mathrm{rc}}(C_n) = \mathrm{RC}(C_n) = n$.
4. $\mathrm{rc}(G) = n$ if and only if G has a hamiltonian path.
5. $\mathrm{RC}(G) \leq n(\Delta + 1)^d$.

The model introduced in [23] is analogous to the robot crawler process, in a way we make precise. For any connected graph $G = (V, E)$ and any $k \in \mathbb{N}$, a *k-subdivision* of G, $L_k(G)$, is a graph that is obtained from G by replacing each edge of G by a path of length k. The following theorem shows the connection between the two models. Recall that $s(G)$ is the analogue of $\mathrm{rc}(G)$ in the robot vacuum model.

Theorem 2. *If $G = (V, E)$ is a connected graph, then*

$$s(G) = \left\lfloor \frac{\mathrm{rc}(L_3(G)) + 1}{3} \right\rfloor.$$

Theorem 2 shows that, indeed, the model we consider in this paper is a generalization of the edge model introduced in [23]. Instead of analyzing $s(G)$ for some connected graph G, we may construct $L_3(G)$ and analyze $\mathrm{rc}(L_3(G))$.

Let us start with the following elementary example to illustrate the robot crawler parameters. For the path P_n of length $n - 1 \geq 2$, we have that $\mathrm{rc}(P_n) = n$ and $\mathrm{RC}(P_n) = 2n - 2$. In order to achieve the minimum, one has to start the process from a leaf of P_n. Regardless of ω_0 used, the process takes n steps to finish (see Lemma 1(4) and Theorem 4 for more general results). In order to achieve the maximum, the robot crawler has to start from a neighbour of a leaf and a weighting that forces the process to move away from the leaf (again, see Theorem 4 for more general result). By direct computation, we have the following result.

Theorem 3. *For any* $n \in \mathbb{N}$,

$$\overline{rc}(P_n) = \frac{3n}{2} - \frac{3}{2} + \frac{1}{n} \sim \frac{3n}{2}.$$

We next give the precise value of rc and RC for trees. The main idea behind the proof of this result is comparing the robot crawler to the Depth-First Search algorithm on a tree.

Theorem 4. *Let* $T = (V, E)$ *be a tree on* $n \geq 2$ *nodes. Then we have that*

$$rc(T) = 2n - 1 - diam(T) \quad and \quad RC(T) = 2n - 2,$$

where $diam(T)$ *is the diameter of* T.

Now, let us move to more sophisticated example. For $k \in \mathbb{N} \setminus \{1\}$ and $n \in \mathbb{N}$, denote the *complete k-partite graph* with partite sets $V_1, \ldots, V_k$ of size n by K_n^k. Note that for any $n \in \mathbb{N}$ and $k = 2$, we have that

$$rc(K_n^2) = \overline{rc}(K_n^2) = RC(K_n^2) = |V(K_n^2)| = 2n.$$

Indeed, since K_n^2 has a hamiltonian path, $rc(K_n^2) = 2n$ (see Lemma 1(4)). However, in fact, regardless of the ω_0 used, the robot crawler starts at a node v_0 and then oscillates between the two partite sets visiting all nodes in increasing order of weights assigned initially to each partite set of K_n^2.

We next consider the case $k \geq 3$. Since K_n^k still has a hamiltonian path, $rc(K_n^k) = kn$. For $RC(K_n^k)$ the situation is slightly more complicated.

Theorem 5. *For any* $k \in \mathbb{N} \setminus \{1, 2\}$ *and* $n \in \mathbb{N}$, *we have that*

$$rc(K_n^k) = kn \text{ and } RC(K_n^k) = (k + 1)n - 1.$$

Investigating $\overline{rc}(K_n^k)$ appears more challenging. However, we derive the asymptotic behaviour.

Theorem 6. *For any* $k \in \mathbb{N} \setminus \{1, 2\}$, *we have that*

$$\overline{rc}(K_n^k) = kn + O(\log n) \sim kn.$$

Before we sketch the proof of Theorem 6, we need a definition. Suppose that we are given an initial weighting ω_0 of K_n^k. For any $\ell \in [kn]$, let A_ℓ be the set of ℓ cleanest nodes; that is,

$$A_\ell = \{v \in V_1 \cup V_2 \cup \ldots \cup V_k : \omega_0(v) \geq -\ell + 1\}.$$

Finally, for any $\ell \in [kn]$ and $j \in [k]$, let $a_\ell^j = a_\ell^j(\omega_0) = |A_\ell \cap V_j|$; that is, a_ℓ^j is the number of nodes of V_j that are among ℓ the cleanest ones (in the whole graph K_n^k). Note that for a random initial weighing ω_0, the expected value of a_ℓ^j is ℓ/k. Let $\varepsilon > 0$. We say that ω_0 is ε-balanced if for each $j \in [k]$ and $6\varepsilon^{-2}k \log n \le \ell \le kn$, we have that

$$\left| a_\ell^j - \frac{\ell}{k} \right| < \frac{\varepsilon\ell}{k}.$$

A crucial observation is that almost all initial weightings are ε-balanced, regardless of how small ε is. We will use the following version of *Chernoff's bound*. Suppose that $X \in \text{Bin}(n, p)$ is a binomial random variable with expectation $\mu = np$. If $0 < \delta < 3/2$, then

$$\Pr\left(|X - \mu| \ge \delta\mu\right) \le 2\exp\left(-\frac{\delta^2\mu}{3}\right). \tag{1}$$

(For example, see Corollary 2.3 in [17].) It is also true that (1) holds for a random variable with the hypergeometric distribution. The *hypergeometric distribution* with parameters N, n, and m (assuming $\max\{n, m\} \le N$) is defined as follows. Let Γ be a set of size n taken uniformly at random from set $[N]$. The random variable X counts the number of elements of Γ that belong to $[m]$; that is, $X = |\Gamma \cap [m]|$. It follows that (1) holds for the hypergeometric distribution with parameters N, n, and m, with expectation $\mu = nm/N$. (See, for example, Theorem 2.10 in [17].)

Now we are ready to state the important lemma which is used in the proof of Theorem 6. Its proof follows from the Chernoff's bound (1) for hypergeometric distributions, and is omitted.

Lemma 2. *Let $\varepsilon > 0$ and $k \in \mathbb{N}\setminus\{1, 2\}$, and let ω_0 be a random initial weighting of K_n^k. Then we have that ω_0 is ε-balanced with probability $1 - O(n^{-1})$.*

Proof of Theorem 6. Let $k \in \mathbb{N} \setminus \{1, 2\}$ and fix $\varepsilon = 0.01$. We will show that for any ε-balanced initial weighting ω_0, $\text{rc}(K_n^k, \omega_0) = kn + O(\log n)$. This will finish the proof since, by Lemma 2, a random initial weighting is ε balanced with probability $1 - O(n^{-1})$, and for any initial weighting ω_0 we have $\text{rc}(K_n^k, \omega_0) \le \text{RC}(K_n^k) = (k+1)n - 1 = O(n)$. Indeed,

$$\overline{\text{rc}}(K_n^k) = \Pr\left(\omega_0 \text{ is } \varepsilon\text{-balanced}\right)(kn + O(\log n)) + \Pr\left(\omega_0 \text{ is not } \varepsilon\text{-balanced}\right)O(n)$$
$$= (kn + O(\log n)) + O(1) = kn + O(\log n).$$

Let ω_0 be any ε-balanced initial weighting. Fix $\ell \in [kn]$ and let us run the process until the robot crawler is about to move for the first time to a node of A_ℓ. Suppose that the robot crawler occupies node $v \in V_i$ for some $i \in [k]$ ($v \notin A_\ell$) and is about to move to node $u \in V_j$ for some $j \in [k], j \ne i$ ($u \in A_\ell$). Let us call V_i a ℓ-*crucial* partite set. Concentrating on non-crucial sets, we observe that for any $s \neq i$, all the nodes of $V_s \setminus A_\ell$ are already cleaned; otherwise, the robot crawler would go to such node, instead of going to u. On the other hand, it might be the case that not all nodes of $V_i \setminus A_\ell$, that belong to a ℓ-crucial set, are already visited; we will call such nodes ℓ-*dangerous*. Let $f(\ell)$ be the number of ℓ-dangerous nodes.

Our goal is to control the function $f(\ell)$. We say that ℓ is *good* if $f(\ell) \leq 0.6\ell/k$. Clearly, $\ell = kn$ is good, as $f(kn) = 0$. We use the following claim.

Claim. If ℓ is good, then $\ell' = \lfloor 2\ell/3 \rfloor$ is good, provided that $\lfloor 2\ell/3 \rfloor \geq 6\varepsilon^{-2}k\log n$.

To show the claim, we run the process and stop at time T_ℓ when the robot crawler is about to move to the fist node of A_ℓ. We concentrate on the time interval from T_ℓ up to time-step $T_{\ell'}$ when a node of $A_{\ell'}$ is about to be cleaned. First, note that during the first phase of this time interval, the crawler oscillates between nodes of $A_\ell \setminus A_{\ell'}$ that are not in the ℓ-crucial set and ℓ-dangerous nodes. Clearly, there are $\ell - \ell' \geq \ell/3$ nodes in $A_\ell \setminus A_{\ell'}$. Since ω_0 is ε-balanced, the number of nodes of the ℓ-crucial set that belong to A_ℓ and $A_{\ell'}$ is at most $(1+\varepsilon)\ell/k$ and at least $(1-\varepsilon)\ell'/k$, respectively. Since

$$\frac{\ell}{3} - \left(\frac{(1+\varepsilon)\ell}{k} - \frac{(1-\varepsilon)\ell'}{k} \right) = \frac{\ell}{3} - \frac{(1+5\varepsilon)\ell}{3k} + O(1) \geq \left(\frac{k-1}{3} - 2\varepsilon \right) \frac{\ell}{k} > 0.64\frac{\ell}{k} \geq f(\ell),$$

this phase lasts $2f(\ell)$ steps and all ℓ-dangerous nodes are cleaned. The claim now follows easily as one can use a trivial bound for the number of ℓ'-dangerous nodes. Regardless which partite set is ℓ'-crucial, since ω_0 is ε-balanced, we can estimate the number of nodes in ℓ'-crucial set that belong to $A_\ell \setminus A'_\ell$. Since ℓ'-dangerous nodes must be in $A_\ell \setminus A'_\ell$, we obtain that

$$f(\ell') \leq \frac{(1+\varepsilon)\ell}{k} - \frac{(1-\varepsilon)\ell'}{k} = \left(\frac{1}{2} + \frac{5}{2}\varepsilon \right) \frac{\ell'}{k} + O(1) < 0.53\frac{\ell'}{k}.$$

It follows that ℓ' is good and the claim holds by induction.

To finish the proof, we keep applying the claim recursively concluding that there exists $\ell < (3/2)6\varepsilon^{-2}k\log n = O(\log n)$ that is good. At time T_ℓ of the process, $\ell + f(\ell) \leq \ell + 0.6\ell/k = O(\log n)$ nodes are still dirty and every other node is visited exactly once. The process ends after at most $2(\ell + f(\ell))$ another steps for the total of at most $kn + (\ell + f(\ell)) = kn + O(\log n)$ steps. $\square$

3 Binomial Random Graphs

The *binomial random graph* $\mathcal{G}(n, p)$ is defined as a random graph with node set $[n]$ in which a pair of nodes appears as an edge with probability p, independently for each pair of nodes. As typical in random graph theory, we consider only asymptotic properties of $\mathcal{G}(n, p)$ as $n \to \infty$, where $p = p(n)$ may and usually does depend on n.

It is known (see, for example, [19]) that a.a.s. $\mathcal{G}(n, p)$ has a hamiltonian cycle (and so also a hamiltonian path) provided that $pn \geq \log n + \log\log n + \omega$, where $\omega = \omega(n)$ is any function tending to infinity together with n. On the other hand, a.a.s. $\mathcal{G}(n,p)$ has no hamiltonian cycle if $pn \leq \log n + \log \log n - \omega$. It is straightforward show that in this case a.a.s. there are more than two nodes of degree at most 1 and so a.a.s. there is no hamiltonian path. Combining these observations, we derive immediately the following result.

Corollary 1. *If $\omega = \omega(n)$ is any function tending to infinity together with n, then the following hold a.a.s.*

1. *If $pn \geq \log n + \log \log n + \omega$, then $\mathrm{rc}(\mathcal{G}(n,p)) = n$.*
2. *If $pn \leq \log n + \log \log n - \omega$, then $\mathrm{rc}(\mathcal{G}(n,p)) > n$.*

The next upper bound on $\mathrm{RC}(\mathcal{G}(n,p))$ follows from Lemma 1(5) and the fact that $\mathcal{G}(n,p)$ has maximum degree at most $n-1$ and a.a.s. diameter 2 for p in the range of discussion.

Corollary 2. *Suppose $pn \geq C\sqrt{n \log n}$, for a sufficiently large constant $C > 0$. Then a.a.s. we have that*
$$\mathrm{RC}(\mathcal{G}(n,p)) \leq n^3.$$

Moreover, we give the following lower bound (whose proof is omitted here).

Theorem 7. *Suppose $C\sqrt{n \log n} \leq pn \leq (1 - \varepsilon)n$, for constants $C > 1$ and $\varepsilon > 0$. Then a.a.s. we hae that*
$$\mathrm{RC}(\mathcal{G}(n,p)) \geq (2 - p + o(p))n.$$

The rest of this section is devoted to the following result.

Theorem 8. *Let $p = p(n)$ such that $pn \gg \sqrt{n \log n}$. Then a.a.s.*
$$\overline{\mathrm{rc}}(\mathcal{G}(n,p)) = n + o(n).$$

The main ingredient to derive Theorem 8 is the following key lemma.

Lemma 3. *Let $G = (V,E) \in \mathcal{G}(n,p)$ for some $p = p(n)$ such that $pn \gg \sqrt{n \log n}$, and let $\omega_0 : V \to B_n$ be any fixed initial weighting. Then with probability $1 - o(n^{-3})$, we have that*
$$\mathrm{rc}(G, \omega_0) = n + o(n).$$

We are going to fix an initial weighting before exposing edges of the random graph. For a given initial weighting $\omega_0 : V \to B_n$, we partition the node set V into 3 types with respect to their initial level of dirtiness: *type 1* consists of nodes with initial weights from $B_n \setminus B_{\lfloor 2n/3 \rfloor}$, *type 2* with initial weights from $B_{\lfloor 2n/3 \rfloor} \setminus B_{\lfloor n/3 \rfloor}$; the remaining nodes are of *type 3*. Before we move to the proof of Lemma 3, we state the following useful claim that holds even for much sparser graphs (the proof is immediate by a standard Chernoff bound (1)).

Claim 1. Let $G = (V, E) \in \mathcal{G}(n, p)$ for some $p = p(n)$ such that $pn \gg \log n$. Let $\omega_0 : V \to B_n$ be any initial weighting. Then the following property holds with probability $1 - o(n^{-3})$. Each node $v \in V$ has $(1 + o(1))pn/3$ neighbours of each of the three types.

We will use the claim in the proof of the main result but not explicitly; that is, we do not want to condition on the property stated in the claim. Instead, we uncover edges of the (unconditional) random graph (one by one, in some order) and show that the desired upper bound for $\mathrm{rc}(\mathcal{G}(n, p), \omega_0)$ holds with the desired probability *unless* the claim is false. Now we can move to the proof of Lemma 3.

Proof of Lemma 3. We consider four phases of the crawling process.

Phase 1: We start the process from the initial node (which is of type 1, since it has initial weight $-n + 1$), and then we clean only nodes of type 1. The phase ends when the robot crawler is not adjacent to any dirty node of type 1; that is, when the crawler is about to move to a node of some other type than type 1 or to re-clean some node of type 1. An important property is that, at any point of the process, potential edges between the crawler and dirty nodes are not exposed yet. Hence, if $x \geq 5 \log n/p$ nodes of type 1 are still dirty, the probability that this phase ends at this point is equal to

$$(1 - p)^x \leq \exp(-px) \leq n^{-5}.$$

Hence, it follows from the union bound that, with probability at least $1 - n^{-4} = 1 - o(n^{-3})$, this phase ends after T_1 steps, where $\lceil n/3 \rceil - 5 \log n/p \leq T_1 \leq \lceil n/3 \rceil$, at most $5 \log n/p$ nodes of type 1 are still dirty, and the other type 1 nodes are cleaned exactly once. Observe that during this phase we exposed only edges between type 1 nodes.

Phase 2: During this phase we are going to clean mostly nodes of type 2, with a few "detours" to type 1 nodes that are still dirty. Formally, the phase ends when the robot crawler is not adjacent to any dirty node of type 1 or 2; that is, when the crawler is about to move to a node of type 3 or to re-clean some node (of type 1 or 2). Arguing as before, we deduce that, with probability at least $1 - o(n^{-3})$, this phase ends after the total of T_2 steps (counted from the beginning of the process), where $\lceil 2n/3 \rceil - 5 \log n/p \leq T_2 \leq \lceil 2n/3 \rceil$, at most $5 \log n/p$ nodes of type 1 or 2 are still dirty, and the other type 1 or 2 nodes are cleaned exactly once.

Suppose that at the end of this phase some node v of type 1 is still dirty. This implies that v has at most $10 \log n/p$ neighbours that are type 2. Indeed, at most $5 \log n/p$ of them are perhaps not visited by the crawler yet; at most $5 \log n/p$ of them were visited by the crawler but it did not move to v from them but went to some other of the at most $5 \log n/p$ dirty nodes of type 1 instead. Since $pn \geq 10\sqrt{n \log n}$, we obtain that $10 \log n/p \leq pn/10$ and so this implies that the property stated in Claim 1 is not satisfied. If this is the case, then we simply stop the argument. We may then assume that all nodes of type 1 are cleaned at this point of the process. Finally, let us mention that during this phase we exposed only edges between type 2 nodes, and between type 1 nodes that were dirty at the end of phase 1 and type 2 nodes.

Phase 3: This phase ends when the robot crawler is not adjacent to any dirty node; that is, when the crawler is about to re-clean some node. During this phase we are going to clean mostly nodes of type 3, with a few "detours" to type 2 nodes that are still dirty. Arguing as before, we deduce that, with probability at least $1 - o(n^{-3})$, this phase ends after the total of T_3 steps, where $n - 5 \log n/p \leq T_2 \leq n$. Moreover, we may assume that at the end of this phase at most $5 \log n/p$ nodes of type 3 are still dirty whereas all other nodes are cleaned exactly once; otherwise, the property stated in Claim 1 is not satisfied. As usual, the main observation is that during this phase we exposed only edges between type 3 nodes, and between type 2 nodes that were dirty at the end of phase 2 and type 3 nodes.

Phase 4: During this final phase we are going to re-clean (for the second time) some nodes of type 1, with a few "detours" to type 3 nodes that are still dirty. This phase ends when one of the following properties is satisfied:

(a) all nodes are cleaned,
(b) this phase takes more than $20 \log n/p^2$ steps,
(c) the robot crawler is not adjacent to any dirty node nor to any type 1 node that was cleaned only once, during phase 1 (note that these nodes have the smallest weights at this point of the process).

Recall that our goal is to show that either the property stated in Claim 1 is not satisfied or, with probability at least $1 - o(n^{-3})$, the phase ends when all nodes are cleaned. From this it will follow that the process takes $n + O(\log n/p^2) = n + o(n)$ steps with probability at least $1 - o(n^{-3})$, and the proof will be finished.

Suppose first that the phase ends because of property (c). It follows that the crawler occupies a node v that has at most $25 \log n/p$ neighbours that are type 1: at most $20 \log n/p$ of them were re-cleaned during this phase, and at most $5 \log n/p$ of them were cleaned during phase 2. Since $pn \geq 10\sqrt{n \log n}$, $25 \log n/p \leq pn/4$ and so the property in Claim 1 is not satisfied. Hence, we may assume that the phase does not end because of (c).

Suppose now that the phase ends because of property (b) and that property (c) is never satisfied. This implies that all nodes visited during phase 4 must be different, since otherwise property (c) would hold. Moreover, the robot crawler can be adjacent to a dirty node at most $5 \log n/p$ out of the first $\lfloor 20 \log n/p^2 \rfloor$ steps in this phase, since each time this happens one dirty node will be cleaned in the next step, and there were at most $5 \log n/p$ nodes of type 3 that were dirty at the end of phase 3. A crucial observation is that no edges between type 1 and type 3 nodes (and also no edges between dirty nodes of type 3) were exposed at the beginning of this phase. Using this we can estimate the probability that at the end of this phase some node is still dirty. Indeed, at each step, the probability that the robot crawler is adjacent to a dirty node (provided that some dirty node still exists) is at least p. Hence, using Chernoff bound (1), the probability that phase 4 ends because of property (b) and not (c) is at most

$$\Pr\left(\text{Bin}(\lfloor 20\log n/p^2\rfloor, p) \le 5\log n/p\right) \le \exp\left(-\frac{(3/4)^2 20\log n/p}{3 + o(1)}\right) = o(n^{-3}).$$

This shows that phase 4 does not stop because of property (b) with probability $1 - o(n^{-3})$, as required.   □

4 Preferential Attachment Model

The results in Sect. 3 demonstrate that for the binomial random graph, for most initial weightings the robot crawler will finish in approximately n steps. We now consider the robot crawler on a stochastic model for complex networks. The *preferential attachment model*, introduced by Barabási and Albert [4], was an early stochastic model of complex networks. We will use the following precise definition of the model, as considered by Bollobás and Riordan in [5] as well as Bollobás, Riordan, Spencer, and Tusnády [6].

Let G_1^0 be the null graph with no nodes (or let G_1^1 be the graph with one node, v_1, and one loop). The random graph process $(G_1^t)_{t\ge 0}$ is defined inductively as follows. Given G_1^{t-1}, we form G_1^t by adding node v_t together with a single edge between v_t and v_i, where i is selected randomly with the following probability distribution:

$$\Pr(i = s) = \begin{cases} \deg(v_s, t-1)/(2t-1) & 1 \le s \le t-1, \\ 1/(2t-1) & s = t, \end{cases}$$

where $\deg(v_s, t-1)$ denotes the degree of v_s in G_1^{t-1}. (In other words, we send an edge e from v_t to a random node v_i, where the probability that a node is chosen as v_i is proportional to its degree at the time, counting e as already contributing one to the degree of v_t.)

For $m \in \mathbb{N} \setminus \{1\}$, the process $(G_m^t)_{t\ge 0}$ is defined similarly with the only difference that m edges are added to G_m^{t-1} to form G_m^t (one at a time), counting previous edges as already contributing to the degree distribution. Equivalently, one can define the process $(G_m^t)_{t\ge 0}$ by considering the process $(G_1^t)_{t\ge 0}$ on a sequence $v_1', v_2', \ldots$ of nodes; the graph G_m^t if formed from G_1^{tm} by identifying nodes $v_1', v_2', \ldots, v_m'$ to form v_1, identifying nodes $v_{m+1}', v_{m+2}', \ldots, v_{2m}'$ to form v_2, and so on. Note that in this model G_m^t is in general a multigraph, possibly with multiple edges between two nodes (if $m \ge 2$) and self-loops. For the purpose of the robot crawler, loops can be ignored and multiple edges between two nodes can be treated as a single edge.

It was shown in [6] that for any $m \in \mathbb{N}$ a.a.s. the degree distribution of G_m^n follows a power law: the number of nodes with degree at least k falls off as $(1 + o(1))ck^{-2}n$ for some explicit constant $c = c(m)$ and large $k \le n^{1/15}$. Let us start with the case $m = 1$, which is easy to deal with, since G_1^n is a forest. Each node sends an edge either to itself or to an earlier node, so the graph consists of components which are trees, each with a loop attached. The expected number of components is then $\sum_{t=1}^{n} 1/(2t-1) \sim (1/2)\log n$ and, since events are independent, we derive that a.a.s. there are $(1/2 + o(1))\log n$ components in G_1^n by Chernoff's bound (1). Moreover, Pittel [26] essentially showed that a.a.s. the largest distance between two nodes in the same component of G_1^n is $(\gamma^{-1} + o(1))\log n$, where γ is the solution of $\gamma e^{1+\gamma} = 1$ (see Theorem 13 in [5]). Hence, the following result holds immediately from Theorem 4.

Theorem 9. *The following properties hold a.a.s. for any connected component* G *of* G_1^n:

$$rc(G) = 2|V(G)| - 1 - diam(G) = 2|V(G)| - O(\log n),$$
$$RC(G) = 2|V(G)| - 2.$$

We may modify slightly the definition of the model to ensure G_1^n is a tree on n nodes, by starting from G_1^2 being an isolated edge and not allowing loops to be created in the process (this is in fact the original model in [4]). For such variant, we would have that a.a.s. $rc(G_1^n) \sim RC(G_1^n) \sim 2n$, as the diameter would be negligible comparing to the order of the graph.

The case $m \geq 2$ is more difficult to investigate. It is known that a.a.s. G_m^n is connected and its diameter is $(1 + o(1))\log n/\log\log n$, as shown in [5], and in contrast to the result for $m = 1$ presented above. We managed to show that for the case $m = 2$, the robot crawler needs substantially more than n steps to clean the graph in this model. This immediately implies (in a strong sense) that G_2^n is not hamiltonian a.a.s.

Theorem 10. *A.a.s.* $rc(G_2^n) \geq (1 + \xi + o(1))n$, *where*

$$\xi = \max_{c \in (0,1/2)} \left(\frac{2\sqrt{c}}{3} - c - \frac{c^2}{6} \right) \approx 0.10919.$$

Proof. Many observations in the argument will be valid for any m but, of course, we will eventually fix $m = 2$. Consider the process $(G_m^t)_{t \geq 0}$ on the sequence of nodes $(v_t)_{t \geq 0}$. We will call node v_i *lonely* if $\deg(v_i, n) = m$; that is, no loop is created at the time v_i is introduced and no other node is connected to v_i later in the process. Moreover, v_i is called *old* if $i \leq cn$ for some constant $c \in (0,1)$ that will be optimized at the end of the argument; otherwise, v_i is called *young*. Finally, v_i is called *j-good* if v_i is lonely and exactly j of its neighbours are old.

Let us begin with the big picture for the case $m = 2$. Suppose that an nodes are young and 1-good, bn nodes are young and 2-good, and dn nodes are old and lonely (which implies that they are 2-good). Clearly, the robot crawler needs to visit all young nodes and all old and lonely ones, which takes at least $(1-c)n + dn$ steps. Observe that each time a young and 2-good node is visited, the crawler must come from an old but not-lonely node and move to another such one right after. Similarly, each time the crawler visits a young and 1-good node, it must come from or move to some node that is old but not lonely. It follows that nodes that are old but not lonely must be visited at least $an/2 + bn + O(1)$ times.

Hence, the process must take at least $(1 - c + d + a/2 + b + o(1))n$ steps, and our hope is that it gives a non-trivial bound for some value of $c \in (0,1)$.

The probability that v_i is lonely is easy to estimate from the equivalent definition of G_m^n obtained in terms of G_1^{mn}. For $i \gg 1$, we derive that

$$\Pr(v_i \text{ is lonely}) = \Pr(\deg(v_i, i) = m) \prod_{t=im+1}^{nm} \left(1 - \frac{m}{2t-1}\right)$$

$$\sim \exp\left(-\sum_{t=im+1}^{nm} \frac{m}{2t-1} + O\left(\sum_{t=im+1}^{nm} t^{-2}\right)\right)$$

$$\sim \exp\left(-\frac{m}{2}\sum_{t=im+1}^{nm} t^{-1}\right) \sim \exp\left(-\frac{m}{2}\log\left(\frac{nm}{im}\right)\right) = \left(\frac{i}{n}\right)^{m/2}.$$

We will also need to understand the behaviour of the following random variable: for $\lfloor cn \rfloor \le t \le n$, let

$$Y_t = \sum_{j \le cn} \deg(v_j, t).$$

In view of the identification between the models G_m^n and G_1^{mn}, it will be useful to investigate the following random variable instead: for $m\lfloor cn \rfloor \le t \le mn$, let

$$X_t = \sum_{j \le cmn} \deg_{G_1^t}(v_j', t).$$

Clearly, $Y_t = X_{tm}$. It follows that $X_{m\lfloor cn \rfloor} = Y_{\lfloor cn \rfloor} = 2m\lfloor cn \rfloor$. Moreover, for $m\lfloor cn \rfloor < t \le mn$,

$$X_t = \begin{cases} X_{t-1} + 1 & \text{with probability } \frac{X_{t-1}}{2t-1}, \\ X_{t-1} & \text{otherwise.} \end{cases}$$

The conditional expectation is given by

$$\mathbb{E}[X_t | X_{t-1}] = (X_{t-1} + 1) \cdot \frac{X_{t-1}}{2t-1} + X_{t-1}\left(1 - \frac{X_{t-1}}{2t-1}\right) = X_{t-1}\left(1 + \frac{1}{2t-1}\right).$$

Taking expectation again, we derive that

$$\mathbb{E}[X_t] = \mathbb{E}[X_{t-1}]\left(1 + \frac{1}{2t-1}\right).$$

Hence, arguing as before, it follows that

$$\mathbb{E}[Y_t] = \mathbb{E}[X_{tm}] = 2m\lfloor cn \rfloor \prod_{s=m\lfloor cn \rfloor+1}^{tm} \left(1 + \frac{1}{2s-1}\right) \sim 2cmn\left(\frac{tm}{cmn}\right)^{1/2} = 2mn\sqrt{ct/n}.$$

Noting that $\mathbb{E}[Y_t] = \Theta(n)$ for any $\lfloor cn \rfloor \le t \le n$, and that Y_t increases by at most m each time (X_t increases by at most one), we obtain that with probability $1 - o(n^{-1})$, $Y_t = \mathbb{E}[Y_t] + O(\sqrt{n \log n}) \sim \mathbb{E}[Y_t]$ (using a standard martingale argument; see Azuma-Hoeffding inequality (see, for example, [17]). Hence, we may assume that $Y_t \sim 2mn\sqrt{ct/n}$ for any $\lfloor cn \rfloor \leq t \leq n$.

The rest of the proof is straightforward. Note that, for a given $t = xn$ with $c \leq x \leq 1$, the probability that an edge generated at this point of the process goes to an old node is asymptotic to $(2mn\sqrt{ct/n})/(2mt) = \sqrt{cn/t} = \sqrt{c/x}$. Moreover, recall that v_t is lonely with probability asymptotic to $(t/n)^{m/2} = x$ for the case $m = 2$. It follows that

$$a \sim \int_c^1 2\sqrt{c/x}(1 - \sqrt{c/x})x\,dx = \frac{4\sqrt{c}}{3} - 2c + \frac{2c^2}{3},$$

$$b \sim \int_c^1 (\sqrt{c/x})^2 x\,dx = c - c^2,$$

$$d \sim \int_0^c x\,dx = \frac{c^2}{2}.$$

Since

$$1 - c + d + a/2 + b \sim 1 + \frac{2\sqrt{c}}{3} - c - \frac{c^2}{6}$$

is maximized at $c = \frac{\left(\left(4+4\sqrt{5}\right)^{2/3}-4\right)^2}{4\left(4+4\sqrt{5}\right)^{2/3}} \approx 0.10380$, the proof follows. $\square$

5   Conclusion and Open Problems

We introduced the robot crawler model, which is a simplified model of web crawling. We studied the minimum, maximum, and average time for the robot crawler process to terminate. We found exact values for these parameters in several graph classes such as trees and complete multi-partite graphs. We have successfully addressed the robot crawler model in binomial random graphs, and considered the rc parameter for preferential attachment graphs in the cases $m = 1, 2$.

Several problems concerning the robot crawler model remain open. We list some of these relevant to our investigation below.

1. Let G_n be the complete k-partite graph with partite sets of sizes $c_1 n, c_2 n,$ $\ldots, c_k n$ for some constants $0 < c_1 \leq c_2 \leq \ldots \leq c_k$. Derive the asymptotic behaviour of $\mathrm{rc}(G_n)$, $\overline{\mathrm{rc}}(G_n)$, and $\mathrm{RC}(G_n)$.
2. Theorem 8 holds for dense random graphs; that is, for $pn \gg \sqrt{n \log n}$. What about sparser random graphs?
3. Can the bound in Corollary 2 be improved? Is it true that $\mathrm{RC}(\mathcal{G}(n, p)) = O(n)$ for a wide range of p? Recall, in view of Theorem 7, that we cannot achieve $\mathrm{RC}(\mathcal{G}(n, p)) = (1 + o(1))n$, provided that $p < 1 - \varepsilon$ for some $\varepsilon > 0$.
4. Properties of the robot crawler remain open in the preferential attachment model when $m > 2$. Fix $m \geq 3$. Is it true that a.a.s. $\mathrm{rc}(G_m^n) \geq (1 + \xi)n$ for some constant $\xi > 0$? Or maybe $\mathrm{rc}(G_m^n) \sim n$? It is possible that there is some threshold m_0 such that for $m \leq m_0$, $\mathrm{rc}(G_m^n) \geq (1 + \xi)n$ for some constant $\xi > 0$ but $\mathrm{rc}(G_m^n) \sim n$ for $m > m_0$.

Our work with the robot crawler is a preliminary investigation. As such, it would be interesting to study the robot crawler process on other models of complex networks, such as random graphs with given expected degree sequence [11], preferential attachment graphs with increasing average degrees [14], or geometric models such as the spatially preferred attachment model [1,12], geographical threshold graphs [9], or GEO-P model [7].

References

1. Aiello, W., Bonato, A., Cooper, C., Janssen, J., Prałat, P.: A spatial web graph model with local influence regions. Internet Math. **5**, 175–196 (2009)
2. Alon, N., Prałat, P., Wormald, N.: Cleaning regular graphs with brushes. SIAM J. Discrete Math. **23**, 233–250 (2008)
3. Applegate, D.L., Bixby, R.E., Chvátal, V., Cook, W.J.: The Traveling Salesman Problem. Princeton University Press, Princeton (2007)
4. Barabási, A.L., Albert, R.: Emergence of scaling in random networks. Science **286**, 509–512 (1999)
5. Bollobás, B., Riordan, O.: The diameter of a scale-free random graph. Combinatorica **24**(1), 5–34 (2004)
6. Bollobás, B., Riordan, O., Spencer, J., Tusnády, G.: The degree sequence of a scale-free random graph process. Random Struct. Algorithms **18**, 279–290 (2001)
7. Bonato, A., Janssen, J., Prałat, P.: Geometric protean graphs. Internet Math. **8**, 2–28 (2012)
8. Bonato, A., Nowakowski, R.J.: The Game of Cops and Robbers on Graphs. American Mathematical Society, Providence (2011)
9. Bradonjić, M., Hagberg, A., Percus, A.: The structure of geographical threshold graphs. Internet Math. **5**, 113–140 (2008)
10. Brin, S., Page, L.: Anatomy of a large-scale hypertextual web search engine. In: Proceedings of the 7th International World Wide Web Conference (1998)
11. Chung, F., Lu, L.: Complex Graphs and Networks. American Mathematical Society, Boston (2006)
12. Cooper, C., Frieze, A., Prałat, P.: Some typical properties of the spatial preferred attachment model. Internet Math. **10**, 27–47 (2014)
13. Cooper, C., Ilcinkas, D., Klasing, R., Kosowski, A.: Derandomizing random walks in undirected graphs using locally fair exploration strategies. Distributed Comput. **24**, 91–99 (2011)
14. Cooper, C., Prałat, P.: Scale free graphs of increasing degree. Random Struct. Algorithms **38**, 396–421 (2011)
15. Gajardo, A., Moreira, A., Goles, E.: Complexity of Langton's ant. Discrete Appl. Math. **117**, 41–50 (2002)
16. Henzinger, M.R.: Algorithmic challenges in web search engines. Internet Math. **1**, 115–126 (2004)
17. Janson, S., Luczak, T., Ruciński, A.: Random Graphs. Wiley, New York (2000)
18. Koenig, S., Szymanski, B., Liu, Y.: Efficient and inefficient ant coverage methods. Ann. Math. Artif. Intell. **31**, 41–76 (2001)
19. Komlós, J., Szemerédi, E.: Limit distribution for the existence of Hamiltonian cycles in a random graph. Discrete Math. **43**(1), 55–63 (1983)
20. Li, Z., Vetta, A.: Bounds on the cleaning times of robot vacuums. Oper. Res. Lett. **38**(1), 69–71 (2010)

21. Malpani, N., Chen, Y., Vaidya, N.H., Welch, J.L.: Distributed token circulation in mobile ad hoc networks. IEEE Trans. Mob. Comput. **4**, 154–165 (2005)
22. Manning, C.D., Raghavan, P., Schütze, H.: Introduction to Information Retrieval. Cambridge University Press, New York (2008)
23. Messinger, M.E., Nowakowski, R.J.: The Robot cleans up. J. Comb. Optim. **18**(4), 350–361 (2009)
24. Messinger, M.E., Nowakowski, R.J., Prałat, P.: Cleaning a network with brushes. Theor. Comput. Sci. **399**, 191–205 (2008)
25. Olston, C., Najork, M.: Web crawling. Found. Trends Inform. Retrieval **4**(3), 175–246 (2010)
26. Pittel, B.: Note on the heights of random recursive trees and random m-ary search trees. Random Struct. Algorithms **5**, 337–347 (1994)
27. Wagner, I.A., Lindenbaum, M., Bruckstein, A.M.: Efficiently searching a graph by a smell-oriented vertex process. Ann. Math. Artif. Intell. **24**, 211–223 (1998)
28. West, D.B.: Introduction to Graph Theory, 2nd edn. Prentice Hall, Upper Saddle River (2001)
29. Yanovski, V., Wagner, I.A., Bruckstein, A.M.: A distributed ant algorithm for efficiently patrolling a network. Algorithmica **37**, 165–186 (2003)

Properties of PageRank on Large Graphs

PageRank in Undirected Random Graphs

Konstantin Avrachenkov[1], Arun Kadavankandy[1]([✉]),
Liudmila Ostroumova Prokhorenkova[2],
and Andrei Raigorodskii[2,3,4,5]

[1] Inria Sophia Antipolis, Valbonne, France
arun.kadavankandy@inria.fr
[2] Yandex, Moscow, Russia
[3] Moscow Institute of Physics and Technology, Moscow, Russia
[4] Moscow State University, Moscow, Russia
[5] Buryat State University, Ulan-ude, Russia

Abstract. PageRank has numerous applications in information retrieval, reputation systems, machine learning, and graph partitioning. In this paper, we study PageRank in undirected random graphs with expansion property. The Chung-Lu random graph represents an example of such graphs. We show that in the limit, as the size of the graph goes to infinity, PageRank can be represented by a mixture of the restart distribution and the vertex degree distribution.

Keywords: PageRank · Undirected random graphs · Expander graphs · Chung-Lu random graphs

1 Introduction

PageRank has numerous applications in information retrieval [20,26,30], reputation systems [19,21], machine learning [3,4], and graph partitioning [1,11]. A large complex network can often be conveniently modeled by a random graph. It is surprising that not many analytical studies are available for PageRank in random graph models. We mention the work [5] where PageRank was analysed in preferential attachment models and the more recent works [9,10] where PageRank was analysed in directed configuration models. According to several studies [16,18,23,29] PageRank and in-degree are strongly correlated in directed networks such as Web graph. Apart from some empirical studies [8,27], to the best of our knowledge, there is no rigorous analysis of PageRank on basic undirected random graph models such as the Erdős–Rényi graph [17] or the Chung-Lu graph [13]. In this paper, we fill this gap and show that in these models PageRank can be represented as a mixture of the restart distribution and the vertex degree distribution when the size of the graph goes to infinity. First, we show the convergence in total variation norm for a general family of random graphs with expansion property. Then, we specialize the results for the Chung-Lu random graph model proving the element-wise convergence. We conclude the paper with numerical experiments and several interesting future research directions.

© Springer International Publishing Switzerland 2015
D.F. Gleich et al. (Eds.): WAW 2015, LNCS 9479, pp. 151–163, 2015.
DOI: 10.1007/978-3-319-26784-5_12

2   Definitions

Let $G^{(n)} = (V^{(n)}, E^{(n)})$ denote a family of random graphs, where $V^{(n)}$ is a vertex set, $|V^{(n)}| = n$, and $E^{(n)}$ is an edge set, $|E^{(n)}| = m$. Denote also by $A^{(n)}$ the associated adjacency matrix with elements

$$A_{ij}^{(n)} = \begin{cases} 1, \text{ if } i \text{ and } j \text{ are connected,} \\ 0, \text{ otherwise.} \end{cases}$$

In this work, we analyze PageRank on undirected graphs and hence $A^T = A$. At the same time, our analysis easily extends to some families of weighted undirected graphs. We omit the superscript index n when it is clear from the context. Let $\underline{1}$ be the vector of ones of an appropriate dimension and let $d = A\underline{1}$ be the vector of (weighted) degrees. It is helpful to define $D = \text{diag}(d)$, a diagonal matrix with the degree sequence on its diagonal. Let $P = AD^{-1}$ be the Markov transition matrix corresponding to the standard random walk on the graph and let $Q = D^{-1/2}AD^{-1/2}$ be the symmetrized transition matrix. In this paper we work with column stochastic matrices. Note that the symmetrized transition matrix is closely related to the normalized Laplacian $\mathcal{L} = I - D^{-1/2}AD^{-1/2} = I - Q$ [12]. Further we will also use the resolvent matrix $R = [I - \alpha P]^{-1}$ and the symmetrized resolvent matrix $S = [I - \alpha Q]^{-1}$.

Note that since Q is a symmetric matrix, its eigenvalues $\lambda_i, i = 1, ..., n$ are real and can be arranged in decreasing order, i.e., $\lambda_1 \geq \lambda_2 \geq ...$. In particular, we have $\lambda_1 = 1$. The value $\delta = 1 - \max\{|\lambda_2|, |\lambda_n|\}$ is called the spectral gap.

In what follows, let K be an arbitrary constant that is not the same everywhere and may change even from one line to the next (of course, not causing any inconsistency).

For two functions $f(n), g(n)$ $g = O(f)$, if $\exists C$, a constant such that $|\frac{g}{f}| \leq C$, for large n, and $g = o(f)$ if $\limsup_{n \to \infty} |\frac{g}{f}| = 0$. Additionally, by $f \gg g$, we mean that $f > Cg$ for any constant C for n large enough.

An event E is said to hold with high probability (w.h.p.) if $\Pr(E) \geq 1 - O(n^{-c})$ for some $c > 0$. Recall that if a finite number of events hold true w.h.p., then so does their intersection. Furthermore, we say that a sequence of random variables in $(\Omega, \mathcal{F}, P)$ $X_n = o(1)$ w.h.p. if there exists a function $\psi(n) = o(1)$ such that the event $\{X_n \leq \psi(n)\}$ holds w.h.p.

In the present work, we consider families of random graphs with the following two properties:

Property I: W.h.p., $d_{max}^{(n)}/d_{min}^{(n)} \leq K$, where $d_{max}^{(n)}$ and $d_{min}^{(n)}$ are the maximum and minimum degrees, respectively.

Property II: W.h.p., $\max\{|\lambda_2^{(n)}|, |\lambda_n^{(n)}|\} = o(1)$.

The above two properties can be regarded as a variation of the expansion property. In the standard case of an expander family, one requires the graphs to be regular and the spectral gap $\delta = 1 - \max\{|\lambda_2|, |\lambda_n|\}$ to be bounded away from zero (see, e.g., [28]). Property I is a relaxation of the regularity condition, whereas Property II is stronger than the requirement for the spectral gap to be bounded away from zero. Properties I and II allow us to consider several standard families of random graphs such as Erdős–Rényi graphs, regular random graphs with increasing average degrees, and Chung-Lu graphs. For Chung-Lu graphs Property I imposes some restriction on the degree spread in the graph. It is worth noting that as a consequence of Property I we consider graphs that are do not have isolated vertices ($d_{min} > 0$) w.h.p.

Recall that the Personalized PageRank vector with a restart distribution vector v is defined as a stationary distribution of the modified Markov chain with the transition matrix

$$\tilde{P} = \alpha P + (1 - \alpha)v\underline{1}^T,$$

where α is a so-called damping factor [20]. We also recall the following useful formula for the Personalized PageRank π when $\alpha < 1$ (see, e.g., [22])

$$\pi = (1 - \alpha)[I - \alpha P]^{-1}v = (1 - \alpha)Rv. \tag{1}$$

3   Convergence in Total Variation

We recall that for two discrete probability distributions u and v, the total variation distance $d_{TV}(u, v)$ is defined as $d_{TV}(u, v) = \frac{1}{2}\sum_i |u_i - v_i|$. This can also be thought of as a 1-norm distance measure in the space of probability vectors, wherein for $x \in \mathbf{R}^n$, 1-norm $||x||_1 = \sum_i |x_i|$, and since for any probability vector π^n, $||\pi^n||_1 = 1 \ \forall n$, it makes sense to talk about convergence in 1-norm or TV-distance. Now we are in a position to formulate the following result.

Theorem 1. *Let a family of graphs $G^{(n)}$ satisfy Properties I and II. If, in addition, $||v||_2 = O(1/\sqrt{n})^1$, the PageRank can be asymptotically approximated in total variation norm by a mixture of the restart distribution v and the vertex degree distribution. Namely, w.h.p.*

$$d_{TV}(\pi^{(n)}, \bar{\pi}^{(n)}) = o(1) \quad as \ n \to \infty,$$

where

$$\bar{\pi}^{(n)} = \frac{\alpha d^{(n)}}{vol(G^{(n)})} + (1 - \alpha)v,$$

and $vol(G^{(n)}) = \sum_i d_i^{(n)}$.

The above expression for the asymptotic PageRank vector is interesting in two respects: first it tells us that the rank vector asymptotically behaves like a convex combination of the stationary vector of a standard random walk with transition matrix P; with the weight being α, and secondly, it starts to resemble the stationary vector as α tends to 1.

---

[1] For a vector $x \in \mathbf{R}^n$, $||x||_2 = \sqrt{\sum_i |x_i|^2}$ is the 2-norm.

Proof. We only consider the case in which $0 < \alpha < 1$, since when $\alpha = 0$ or $\alpha = 1$, the statement of the theorem holds trivially.

We first note that the matrix Q can be written as follows by Spectral Decomposition Theorem [6]:

$$Q = u_1 u_1^T + \sum_{i=2}^{n} \lambda_i u_i u_i^T, \tag{2}$$

where $1 = \lambda_1 \geq \lambda_2 \geq \ldots \geq \lambda_n$ are the eigenvalues and $\{u_1, u_2, \ldots u_n\}$ are the corresponding orthogonal eigenvectors, $u_i \in \mathbf{R}^n$, $||u_i||_2 = 1$, and $u_1 = D^{1/2}\mathbf{1}/\sqrt{\mathbf{1}^T D\mathbf{1}}$ is the Perron–Frobenius eigenvector. Next, we rewrite (1) in terms of the matrix Q

$$\pi = (1 - \alpha)D^{1/2}[I - \alpha Q]^{-1}D^{-1/2}v.$$

Substituting (2) into the above equation, we obtain

$$\pi = (1 - \alpha)D^{1/2}\left(\frac{1}{1 - \alpha}u_1 u_1^T + \sum_{i=2}^{n}\frac{1}{1 - \alpha\lambda_i}u_i u_i^T\right)D^{-1/2}v$$

$$= D^{1/2}u_1 u_1^T D^{-1/2}v + (1 - \alpha)D^{1/2}\left(\sum_{i\neq 1}\frac{1}{1 - \alpha\lambda_i}u_i u_i^T\right)D^{-1/2}v.$$

Let us denote the error by $\epsilon = \pi - \bar{\pi}$. Then, we can write

$$\epsilon = \pi - \alpha D^{1/2}u_1 u_1^T D^{-1/2}v - (1 - \alpha)D^{1/2}ID^{-1/2}v$$

$$= (1 - \alpha)D^{1/2}\left(\sum_{i\neq 1}\frac{u_i u_i^T}{1 - \alpha\lambda_i} - (I - u_1 u_1^T)\right)D^{-1/2}v$$

$$= (1 - \alpha)D^{1/2}\left(\sum_{i\neq 1}u_i u_i^T\frac{\alpha\lambda_i}{1 - \alpha\lambda_i}\right)D^{-1/2}v.$$

Now let us bound the 1-norm $||\epsilon||_1$ of the error:

$$||\epsilon||_1/(1 - \alpha) = \left\|D^{1/2}\left(\sum_{i\neq 1}u_i u_i^T\frac{\alpha\lambda_i}{1 - \alpha\lambda_i}\right)D^{-1/2}v\right\|_1$$

$$\leq C\sqrt{d_{max}/d_{min}}\sqrt{n}\max(|\lambda_2|, |\lambda_n|)\,||v||_2 \tag{3}$$

by using $\frac{||Ax||_1}{||x||_2} \leq \sqrt{n}\frac{||Ax||_2}{||x||_2} \leq \sqrt{n}\,||A||_2$,[2] for any A and x, the submultiplicative property of matrix norm[3] and the fact that $||A||_2 = \max_i |\lambda_i|$ if A is Hermitian [6]. Hence we have that $||\epsilon||_1 = o(1)$ w.h.p., under Properties I and II, when $||v||_2 = O(1/\sqrt{n})$. □

---

[2] For any matrix $A \in \mathbf{R}^{m,n}$, $||A||_2 = \sup_{x,||x||_2=1}||Ax||_2$ [6].

[3] For two matrices $A \in \mathbf{R}^{m,n}$, and $B \in \mathbf{R}^{n,p}$, $||AB||_2 \leq ||A||_2\,||B||_2$.

Note that in the case of the standard PageRank, $v_i = 1/n$ implies $\|v\|_2 = O(1/\sqrt{n})$, but Theorem 1 also admits more general restart distributions than the uniform one.

Corollary 1. *The statement of Theorem 1 also holds with respect to the weak convergence, i.e., for any function f on V such that $\max_{x \in V} |f(x)| \leq 1$,*

$$\sup \left\{ \sum_v f(v)\pi_v - \sum_v f(v)\bar{\pi}_v \right\} = o(1) \quad w.h.p.$$

Proof. This follows from Theorem 1 and the fact that the left-hand side of the above equation is upper bounded by $2\,d_{\mathrm{TV}}(\pi_n, \bar{\pi}_n)$ [24]. □

4 Chung-Lu Random Graphs

In this section, we analyze the asymptotics of the PageRank vector in random graphs. As a model of random graphs we consider the Chung-Lu model [13], which is a generalization of the Erdős–Rényi graph model, and hence our results naturally hold for Erdős–Rényi graphs also. The spectral properties of Chung-Lu graphs have been studied extensively in a series of papers [14,15].

4.1 Chung-Lu Random Graph Model

Let us first provide a definition of the Chung-Lu random graph model.

Definition 1 (Chung-Lu Random Graph Model). *A Chung-Lu graph $\mathcal{G}(w)$ with an expected degree vector $w = (w_1, w_2, \ldots w_n)$, where w_i are positive real numbers, is generated by drawing an edge between any two vertices v_i and v_j independently of all others with probability $p_{ij} = \frac{w_i w_j}{\sum_k w_k}$, with the condition of existence $\max_i w_i^2 \leq \sum_k w_k$.*

Below we specify a corollary of Theorem 1 as applied to these graphs. But before that we need the following lemmas about Chung-Lu graphs mainly taken from [14,15].

Lemma 1. *If the expected degrees $w_1, w_2, \ldots w_n$ satisfy $w_{min} \gg \log(n)$, then in $\mathcal{G}(w)$ we have, w.h.p., $\max_i |\frac{d_i}{w_i} - 1| = o(1)$.*

This result is shown in the sense of convergence in probability in [15], but it can be extended to the above result using Bernstein Concentration Lemma [7]:

$$\Pr\{|Y_n - \mathrm{E}Y_n| \geq \epsilon\} \leq 2 \exp \frac{-\epsilon^2}{2(B_n^2 + b\epsilon/3)},$$

where $B_n^2 = \mathrm{E}(Y_n - \mathrm{E}Y_n)^2$, $S_n = X_1 + \ldots X_n$, X_i are independent, and $|X_i| \leq b$. Applying this lemma to the degrees d_i of the Chung-Lu graph we see that

$$\Pr\left(\max_{1 \leq i \leq n} \left|\frac{d_i}{w_i} - 1\right| \geq \beta\right) \leq \frac{2}{n^{c/4-1}}, \quad \text{if} \quad \beta \geq \sqrt{\frac{c\log(n)}{w_{min}}} = o(1)$$

if $w_{min} \gg \log(n)$.

Lemma 2. If $w_{max} \leq K w_{min}$, and $\bar{w} = \sum_k w_k/n \gg \log^6(n)$, then for $\mathcal{G}(w)$ we have almost surely

$$\|C\|_2 = \frac{2}{\sqrt{\bar{w}}}(1 + o(1)),$$

where $C = W^{-1/2}AW^{-1/2} - \chi^T\chi$, $W = diag(w)$, and $\chi_i = \sqrt{w_i/\sum_k w_k}$ is a row vector.

This lemma is an application of Theorem 5 in [14]. It can be verified that when $w_{max} \leq K w_{min}$ and $\bar{w} \gg \log^6(n)$, the condition in Theorem 5 is satisfied, namely, $w_{min} \gg \sqrt{\bar{w}} \log^3(n)$, and hence the result follows.

Lemma 3. For $\mathcal{G}(w)$ with $w_{max} \leq K w_{min}$, and $\bar{w} \gg \log^6(n)$,

$$\max(\lambda_2(P), -\lambda_n(P)) = o(1) \quad w.h.p.,$$

where P is the Markov matrix.

Proof. We have $\|Q - W^{-1/2}AW^{-1/2}\| = o(1)$ w.h.p. using Lemma 1 and the same argument as in the last part of the proof of Theorem 2 in [15]. From this, using Bauer-Fike Lemma [6], we get $|\lambda_i(P) - \lambda_i(W^{-1/2}AW^{-1/2})| = o(1)$ w.h.p. Then, using Lemma 3, we conclude that $\max_{i=1,2,...n} |\lambda_i(C)| = o(1)$ for $\bar{w} \gg O(\log^6(n))$, as a consequence of Bauer-Fike Lemma and the fact that $\chi^T\chi$ is unit rank. So, $\max_{i=1,2...n} |\lambda_i(P)| \leq \max_{i=1,2..n} |\lambda_i(C)| + |\lambda_i(P) - \lambda_i(C)| = o(1)$ w.h.p. $\square$

Corollary 2. Let $\|v\|_2 = O(1/\sqrt{n})$, and $\alpha < 1$. Then PageRank π of the Chung-Lu graph can asymptotically be approximated in TV distance by $\bar{\pi}$, if $\bar{w} \gg \log^6(n)$ and $w_{max} \leq K w_{min}$ for some K that does not depend on n.

Proof. Using Lemma 1 and the condition that $w_{max} \leq K w_{min}$, one can show that $\exists K'$ s.t. $\frac{d_{max}}{d_{min}} \leq K'$ w.h.p. Then the result is a direct consequence of Lemma 3 and the inequality from (3). $\square$

We further note that this result holds also for Erdős-Rényi graphs $\mathcal{G}(n, p)$ with $np_n \gg \log^6(n)$, where we have $(w_1, w_2, \ldots w_n) = (np_n, np_n, \ldots np_n)$.

4.2   Element-Wise Convergence

Earlier in this section, we proved the convergence of PageRank in TV distance for Chung-Lu random graphs. Note that since each component of PageRank could decay to zero as the graph size grows to infinity, this does not necessarily guarantee a convergence in an element-wise sense. In this section, we provide a proof for our convergence conjecture to include the element-wise convergence of the PageRank vector. Here we deviate slightly from the spectral decomposition technique and eigenvalue bounds used hitherto, and instead rely on well-known concentration bounds to bound the error in convergence.

Let $\bar{\Pi} = diag\{\bar{\pi}_1, \bar{\pi}_2, \ldots \bar{\pi}_n\}$ be a diagonal matrix whose diagonal elements are made of the components of the approximated PageRank vector and $\tilde{\delta} = \bar{\Pi}^{-1}(\pi - \bar{\pi})$, i.e., $\tilde{\delta}_i = (\pi_i - \bar{\pi}_i)/\bar{\pi}_i$. Then

$$\tilde{\delta}_i = \left((1-\alpha)v_i + \alpha \frac{d_i}{\text{vol(G)}} \right)^{-1} \left[D^{1/2} \sum_{j\neq 1} \frac{\alpha\lambda_j}{1-\alpha\lambda_j} u_j u_j^T D^{-1/2}v \right]_i.$$

Therefore,

$$\left\| \tilde{\delta} \right\|_\infty \leq \frac{\sum_i d_i/n}{\alpha d_{\min}} \sqrt{d_{max}} \left\| \sum_{i\neq 1} \frac{\alpha\lambda_i}{1-\alpha\lambda_i} u_i u_i^T \tilde{v}' \right\|_\infty, \tag{4}$$

where $\tilde{v}' \equiv nD^{-1/2}v$, and $\left\| \tilde{\delta} \right\|_\infty = \max_i |\tilde{\delta}_i|$.

Define $\tilde{Q} = Q - u_1 u_1^T$, the restriction of the matrix Q to the orthogonal subspace of u_1. Later in this section we prove the following lemma.

Lemma 4. *For a Chung-Lu random graph $\mathcal{G}(w)$ with expected degrees w_1, $w_2, \ldots w_n$, where $w_{max} \leq Kw_{min}$ and $w_{min} \gg \log(n)$, we have with high probability, $\left\| \tilde{Q}\tilde{v}' \right\|_\infty = o(1/\sqrt{w_{min}})$, when $v_i = O(1/n) \; \forall i$.*

The next lemma is related to the matrix $S = (I - \alpha Q)^{-1}$, as defined earlier in the paper.

Lemma 5. *Under the conditions of Lemma 4, $\|S\|_\infty \leq C$ w.h.p., where C is a number independent of n that depends only on α and K.*

Proof. Note that $S = (I - \alpha Q)^{-1} = D^{-1/2}(I - \alpha P)^{-1}D^{1/2}$. Therefore, $\|S\|_\infty \leq \sqrt{\frac{d_{max}}{d_{min}}} \left\| (I - \alpha P)^{-1} \right\|_\infty$ and the result follows since $\left\| (I - \alpha P)^{-1} \right\|_\infty \leq \frac{1}{1-\alpha}$ [22] and using Lemma 1. □

Theorem 2. *Let $v_i = O(1/n) \; \forall i$, and $\alpha < 1$. PageRank π converges element-wise to $\bar{\pi} = (1-\alpha)v + \alpha d/vol(G)$, in the sense that $\max_i (\pi_i - \bar{\pi}_i)/\bar{\pi}_i = o(1)$ w.h.p., on the Chung-Lu graph with expected degrees $\{w_1, w_2, \ldots w_n\}$ such that $w_{min} > \log^c(n)$ for some $c > 1$ and $w_{max} \leq Kw_{min}$, K being a constant independent of n.*

Proof. Define $Z := \sum_{i\neq 1} \frac{\alpha\lambda_i}{1-\alpha\lambda_i} u_i u_i^T$. We then have:

$$Z = \sum_{i=1}^n \frac{\alpha\lambda_i}{1-\alpha\lambda_i} u_i u_i^T - \frac{\alpha}{1-\alpha} u_1 u_1^T = (I - \alpha Q)^{-1}\alpha Q - \frac{\alpha}{1-\alpha} u_1 u_1^T$$

$$= S \left[\alpha Q - \frac{\alpha}{1-\alpha}(I - \alpha Q)u_1 u_1^T \right] = \alpha S\tilde{Q}$$

Using Lemmas 4 and 5, and Eq. (4), we have:

$$\left\|\tilde{\delta}\right\|_\infty = C\frac{\sum_i d_i/n}{d_{min}}\sqrt{d_{max}}\, o(1/\sqrt{w_{min}}) = C\left(\frac{d_{max}}{d_{min}}\right)^{\frac{3}{2}} o(1), \tag{5}$$

which using Lemma 1 is $o(1)$ w.h.p.   □

Corollary 3 (Erdős–Rényi graphs). *For an Erdős-Rényi graph $G(n,p)$ with $np_n \gg \log(n)$, we have that asymptotically the personalized PageRank π converges pointwise to $\bar{\pi}$ for v such that $v_i = O(1/n)$.*

Proof for Lemma 4: From Lemma 1, we have for Chung-Lu graphs that: $d_i = w_i(1 + \epsilon_i)$, where $\omega \equiv \max_i \epsilon_i = o(1)$ with high probability. In the proof we assume explicitly that $v_i = 1/n$, but the results hold in the slightly more general case where $v_i = O(1/n)$ uniformly $\forall i$, i.e., $\exists K$ such that $\max_i nv_i \leq K$. It can be verified easily that all the bounds that follow hold in this more general setting. The event $\{\omega = o(1)\}$, holds w.h.p. asymptotically from Lemma 1. In this case, we have

$$\sum_j \left(\frac{A_{ij}}{\sqrt{d_i d_j}} - \frac{\sqrt{d_i d_j}}{\sum_i d_i}\right)\frac{v_j}{\sqrt{d_j}} = \sum_j \left(\frac{A_{ij}}{\sqrt{d_i d_j}} - \frac{\sqrt{d_i d_j}}{\sum_k d_k}\right)\frac{v_j}{\sqrt{w_j}}(1+\varepsilon_j)$$

where ε_j is the error of convergence, and we have $\max_j \varepsilon_j = O(\omega)$. Therefore,

$$\left\|\tilde{Q}\tilde{v}'\right\|_\infty \leq \left\|\tilde{Q}q\right\|_\infty + \max_i \varepsilon_i \left\|\tilde{Q}q\right\|_\infty \leq \left\|\tilde{Q}q\right\|_\infty (1+o(1)) \quad \text{w.h.p.},$$

where q is a vector such that $q_i = \frac{nv_i}{\sqrt{w_i}}$. Furthermore, we have w.h.p.

$$\frac{A_{ij}}{\sqrt{d_i d_j}} - \frac{\sqrt{d_i d_j}}{\sum_k d_k} = \frac{A_{ij}}{\sqrt{w_i(1+\epsilon_i)w_j(1+\epsilon_j)}} - \frac{\sqrt{w_i(1+\epsilon_i)w_j(1+\epsilon_j)}}{\sum_k w_k(1+\epsilon_k)}$$

$$= \frac{A_{ij}}{\sqrt{w_i w_j}}(1 + O(\epsilon_i) + O(\epsilon_j)) - \frac{\sqrt{w_i w_j}}{\sum_k w_k}\left(\frac{1 + O(\epsilon_i) + O(\epsilon_j)}{1 + O(\omega)}\right)$$

$$= \left(\frac{A_{ij}}{\sqrt{w_i w_j}} - \frac{\sqrt{w_i w_j}}{\sum_k w_k}\right)(1 + \delta_{ij}),$$

where δ_{ij} is the error in the ij^{th} term of the matrix and $\delta_{ij} = O(\omega)$ uniformly, so that $\max_{ij}\delta_{ij} = o(1)$ w.h.p. Consequently, defining $\overline{\overline{Q}}_{ij} = \frac{A_{ij}}{\sqrt{w_i w_j}} - \frac{\sqrt{w_i w_j}}{\sum_k w_k}$ we have:

$$\left\|\tilde{Q}q\right\|_\infty \leq \left\|\overline{\overline{Q}}q\right\|_\infty + \max_i |\sum_j \overline{\overline{Q}}_{ij}\delta_{ij}q_j|$$

$$\leq \left\|\overline{\overline{Q}}q\right\|_\infty + O(\omega)\max_i \frac{1}{\sqrt{w_{min}}}\sum_j |\overline{\overline{Q}}_{ij}|$$

$$\leq \left\|\overline{\overline{Q}}q\right\|_\infty + o(1)\frac{1}{\sqrt{w_{min}}}\left(\sqrt{\frac{w_{max}}{w_{min}}} + \frac{w_{max}}{w_{min}}\right) \tag{6}$$

$$\leq \left\|\overline{\overline{Q}}q\right\|_\infty + o(1/\sqrt{w_{min}}) \tag{7}$$

where in (6) we used the fact the $O(\omega)$ is a uniform bound on the error and it is $o(1)$, and the fact that $\max_i \sum_j |\tilde{Q}_{ij}| \leq \max_i \sum_j \frac{A_{ij}}{\sqrt{w_i w_j}} + \sum_j \frac{\sqrt{w_i w_j}}{\sum_k w_k} \leq \sqrt{\frac{w_{max}}{w_{min}}} + \frac{w_{max}}{w_{min}}$ from simple bounds, and $\max_j q_j \leq \frac{1}{\sqrt{w_{min}}}$. Now we proceed to bound $\left\| \tilde{Q} q \right\|_\infty$. Substituting for $q_i = \frac{1}{\sqrt{w_i}}$, we get

$$\sum_j \frac{1}{\sqrt{w_j}} \left(\frac{A_{ij}}{\sqrt{w_i w_j}} - \frac{\sqrt{w_i w_j}}{\sum_k w_k} \right) = \sum_j \frac{1}{w_j \sqrt{w_i}} \left(A_{ij} - \frac{w_i w_j}{\sum_i w_i} \right) \equiv \frac{1}{\sqrt{w_i}} X_i. \quad (8)$$

We seek to bound $\max_i |X_i| : X_i = \sum_j \frac{1}{w_j} \left(A_{ij} - \frac{w_i w_j}{\sum_i w_i} \right) \equiv Y_n - EY_n$, where $Y_n = \sum_j \frac{A_{ij}}{w_j}$. Furthermore, $EX_n^2 = \sum_j \frac{1}{w_j^2} E(A_{ij} - p_i)^2$, with $p_i = \frac{w_i w_j}{\sum_i w_i}$. So, $ES_n^2 \leq \frac{w_i}{\sum_i w_i} \sum_j \frac{1}{w_j} \leq n \frac{p_i}{w_{min}}$, and $\frac{A_{ij}}{w_j} \leq 1/w_{min}$. Therefore by use of Bernstein Concentration Lemma for $\epsilon < n \max_i p_i$:

$$\Pr \left\{ \max_i | \sum_j (A_{ij} - p_{ij})/w_j | \geq \epsilon \right\} \leq n \max_i \exp(-\frac{\epsilon^2}{2(p_i n/w_{min}) + \epsilon/w_{min}})$$

$$\leq n \max_i \exp(-\frac{w_{min} \epsilon^2}{2(np_i + \epsilon)}) \leq n \exp(-\epsilon^2 w_{min}/(4n \max_i p_i))$$

$$\leq n \exp(\frac{-\epsilon^2 \text{vol} w_{min}}{4 w_{max} n}),$$

where $\frac{\text{vol}}{n} = \frac{\sum_i w_i}{n} \geq w_{min}$. It can be verified that when $\epsilon = \frac{1}{(\bar{w})^\alpha}$ for example, for some $\alpha > 0$, the probability above can be upper bounded by $n^{-(\gamma K - 1)}$, if $\bar{w} \geq (\gamma \log(n))^{\frac{1}{1-2\alpha}}$, for some γ large enough, which can be easily satisfied if $w_{min} \gg O(\log^c(n))$, for some $c > 1$. Thus, finally, from (8) and (7) we have $\left\| \tilde{Q} q \right\|_\infty = o(1/\sqrt{w_{min}})$, w.h.p., and therefore from (4.2), we get $\left\| \tilde{Q} \tilde{v}' \right\|_\infty = o(1/\sqrt{w_{min}})$. $\square$

5   Experimental Results

In this section, we provide experimental evidence to further illustrate the analytic results obtained in the previous sections. In particular, we simulated Erdős–Rényi graphs with $p_n = C \frac{\log^7(n)}{n}$ and Chung-Lu graphs with the degree vector w sampled from a geometric distribution so that the average degree $\bar{w} = O(n^{1/3})$, clipped such that $w_{max} = 7 w_{min}$, for various values of graph size, and plotted the maximum of relative error $\bar{\delta}$ and TV distance error $\|\delta\|_1$, respectively, in Figs. 1 and 2. As expected, both these errors decay as functions of the graph size, which illustrates that the PageRank vector does converge to the asymptotic value. In the interest of exploration, we also conducted simulations on power-law graphs with exponent 5, and it appears that PageRank converges for these graphs as well, albeit the decay of the error being slightly noisier than observed in the previous examples (see Fig. 3). This requires further study.

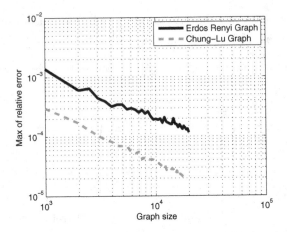

Fig. 1. Log-log plot of maximum relative error for Erdős–Rényi and Chung-Lu graphs

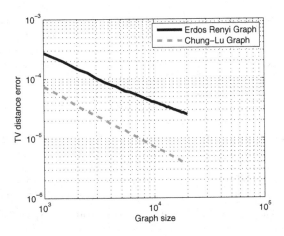

Fig. 2. Log-log plot of TV distance error for Erdős–Rényi and Chung-Lu graphs

Furthermore, we see that when $v_i = 1$ for some i the convergence does not hold (Fig. 4 in the case of Erdős–Rényi graphs). Whereas we see from our analysis that if $v_k = 1$ for some k, the quantity $\left\| \widetilde{Q} D^{-1/2} v \right\|_\infty$, becomes:

$$\max_i \left| \sum_j \left(\frac{A_{ij}}{\sqrt{d_i d_j}} - \frac{\sqrt{d_i d_j}}{\sum_l d_l} \right) v_j / \sqrt{d_j} \right| = \max_i \frac{1}{\sqrt{d_i d_k}} \left| A_{ik} - \frac{d_i d_k}{\sum_l d_l} \right|,$$

which is $O\left(\frac{1}{\sqrt{w_{min} w_k}} \right)$ and does not fall sufficiently fast.

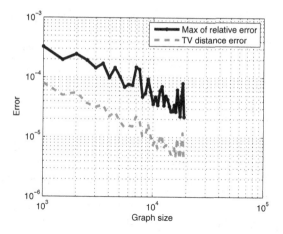

Fig. 3. Log-log plot of TV distance and maximum relative error for power-law graphs

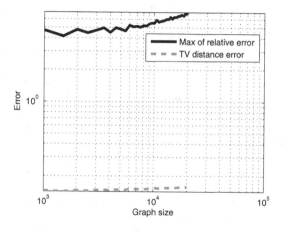

Fig. 4. Log-log plot of TV distance and maximum relative error for ER-graph when $v = e_1$

6   Conclusions

In this work, we showed that when the size of a graph tends to infinity, the PageRank vector lends itself to be approximated by a mixture of the restart distribution and the degree distribution, subject to some conditions for a class of random graphs. In future, we would like to relax some of these conditions, especially the condition on degree spread. This condition can be relaxed as demonstrated by simulations on power-law graphs. In addition, we would like to obtain closer bounds on the error for Chung-Lu graphs as empirical evidence suggests these can be improved too.

Acknowledgements. We would like to thank Nelly Litvak for stimulating discussions on the topic of the paper. This work was partially supported by the French Government (National Research Agency, ANR) through the "Investments for the Future" Program reference ANR-11-LABX-0031-01 and ADR "Network Science" from Joint Inria Alcatel-Lucent Lab.

References

1. Andersen, R., Chung, F., Lang, K.: Local graph partitioning using pagerank vectors. In: Proceedings of IEEE FOCS (2006)
2. Avrachenkov, K., Cottatellucci, L., Kadavankandy, A.: Spectral properties of random matrices for stochastic block model. In Proceedings of WiOpt Workshop PhysComNet (2015)
3. Avrachenkov, K., Dobrynin, V., Nemirovsky, D., Pham, S.K., Smirnova, E.: Pagerank based clustering of hypertext document collections. In: Proceedings of ACM SIGIR, pp. 873–874 (2008)
4. Avrachenkov, K., Gonçalves, P., Mishenin, A., Sokol, M.: Generalized optimization framework for graph-based semi-supervised learning. In: Proceedings of SIAM Conference on Data Mining, vol. 9 (2012)
5. Avrachenkov, K., Lebedev, D.: PageRank of scale-free growing networks. Internet Math. **3**(2), 207–231 (2006)
6. Bhatia, R.: Matrix analysis. Springer Sci. Bus. Media **169**, (2013)
7. Billingsley, P.: Probability and measure. Wiley, New York (2008)
8. Boudin, F.: A comparison of centrality measures for graph-based keyphrase extraction. In: Proceedings of the International Joint Conference on Natural Language Processing (IJCNLP) (2013)
9. Chen, N., Litvak, N., Olvera-Cravioto, M.: Pagerank in scale-free random graphs. In: Bonato, A., Graham, F.C., Prałat, P. (eds.) WAW 2014. LNCS, vol. 8882, pp. 120–131. Springer, Heidelberg (2014)
10. Chen, N., Litvak, N., Olvera-Cravioto, M.: Ranking algorithms on directed configuration networks (2014). Preprint arXiv:1409.7443
11. Chung, F.: A local graph partitioning algorithm using heat kernel pagerank. In: Avrachenkov, K., Donato, D., Litvak, N. (eds.) WAW 2009. LNCS, vol. 5427, pp. 62–75. Springer, Heidelberg (2009)
12. Chung, F.: Spectral Graph Theory. American Mathematical Soc, Providence (1997)
13. Chung, F., Lu, L.: The average distances in random graphs with given expected degrees. PNAS **99**(25), 15879–15882 (2002)
14. Chung, F., Lu, L., Vu, V.: Spectra of random graphs with given expected degrees. PNAS **100**(11), 6313–6318 (2003)
15. Chung, F., Radcliffe, M.: On the spectra of general random graphs. Electron. J. Comb. **18**(1), 215 (2011)
16. Ding, C., He, X., Husbands, P., Zha, H., Simon, H.D.: PageRank. In: Proceedings of ACM SIGIR HITS and a Unified Framework for Link Analysis (2002)
17. Erdős, P., Rényi, A.: On random graphs. Publicationes Math. Debrecen **6**, 290–297 (1959)
18. Fortunato, S., Boguñá, M., Flammini, A., Menczer, F.: Approximating pagerank from in-degree. In: Aiello, W., Broder, A., Janssen, J., Milios, E.E. (eds.) WAW 2006. LNCS, vol. 4936, pp. 59–71. Springer, Heidelberg (2008)

19. Gkorou, D., Vinko, T., Pouwelse, J., Epema, D.: Leveraging node properties in random walks for robust reputations in decentralized networks. In: Proceedings of IEEE Peer-to-Peer Computing (P2P) (2013)
20. Haveliwala, T.H.: Topic-sensitive pagerank. In: Proceedings of WWW, pp. 517–526 (2002)
21. Kamvar, S.D., Schlosser, M.T., Garcia-Molina, H.: The eigentrust algorithm for reputation management in p2p networks. In: Proceedings of WWW (2003)
22. Langville, A.N., Meyer, C.D.: Deeper inside pagerank. Internet Math. 1(3), 335–380 (2004)
23. Litvak, N., Scheinhardt, W.R., Volkovich, Y.: In-degree and pagerank: why do they follow similar power laws? Internet Math. 4(2–3), 175–198 (2007)
24. Levin, D.A., Peres, Y., Wilmer, E.L.: Markov chains and mixing times. American Mathematical Soc, Providence (2009)
25. Moler, C., Moler, K.: Numerical Computing with MATLAB. SIAM (2003)
26. Page, L., Brin, S., Motwani, R., Winograd, T.: Pagerank: bringing order to the web. Stanford Digital Libraries Working Paper, v. 72 (1997)
27. Perra, N., Fortunato, S.: Spectral centrality measures in complex networks. Phys. Rev. E 78, 036107 (2008)
28. Vadhan, S.: Pseudorandomness. Found. Trends Theoret. Comput. Sci. 7(1–3), 1–336 (2012)
29. Volkovich, Y., Litvak, N.: Asymptotic analysis for personalized web search. Adv. Appl. Prob. 42(2), 577–604 (2010)
30. Yeh, E., Ramage, D., Manning, C.D., Agirre, E., Soroa, A.: WikiWalk: random walks on Wikipedia for semantic relatedness. In: Proceedings of the Workshop on Graph-based Methods for Natural Language Processing (2009)

Bidirectional PageRank Estimation: From Average-Case to Worst-Case

Peter Lofgren[1], Siddhartha Banerjee[2]([✉]), and Ashish Goel[1]

[1] Stanford University, Stanford, CA 94305, USA
[2] Cornell University, Ithaca, NY 14850, USA
sbanerjee@cornell.edu

Abstract. We present a new algorithm for estimating the Personalized PageRank (PPR) between a source and target node on undirected graphs, with sublinear running-time guarantees over the worst-case choice of source and target nodes. Our work builds on a recent line of work on bidirectional estimators for PPR, which obtained sublinear running-time guarantees but in an average-case sense, for a uniformly random choice of target node. Crucially, we show how the reversibility of random walks on undirected networks can be exploited to convert average-case to worst-case guarantees. While past bidirectional methods combine forward random walks with reverse local pushes, our algorithm combines forward local pushes with reverse random walks. We also discuss how to modify our methods to estimate random-walk probabilities for any length distribution, thereby obtaining fast algorithms for estimating general graph diffusions, including the heat kernel, on undirected networks.

1 Introduction

Ever since their introduction in the seminal work of Page et al. [23], PageRank and Personalized PageRank (PPR) have become some of the most important and widely used network centrality metrics (a recent survey [13] lists several examples). At a high level, for any graph G, given 'teleport' probability α and a 'personalization distribution' σ over the nodes of G, PPR models the importance of every node from the point of view of σ in terms of the stationary probabilities of 'short' random walks that periodically restart from σ with probability α. It can be defined recursively as giving importance α to σ, and in addition giving every node importance based on the importance of its in-neighbors.

Formally, given normalized adjacency matrix $W = D^{-1}A$, the Personalized PageRank vector π_σ with respect to source distribution σ is the solution to

$$\pi_\sigma = \alpha\sigma + (1 - \alpha)\pi_\sigma W. \tag{1}$$

An equivalent definition is in terms of the terminal node of a random-walk starting from σ. Let $\{X_0, X_1, X_2, \ldots\}$ be a random-walk starting from $X_0 \sim \sigma$, and $L \sim Geometric(\alpha)$. Then the PPR of any node t is given by [4]:

$$\pi_\sigma(t) = \mathbb{P}[X_L = t] \tag{2}$$

The equivalence of these definitions can be seen using a power series expansion.

© Springer International Publishing Switzerland 2015
D.F. Gleich et al. (Eds.): WAW 2015, LNCS 9479, pp. 164–176, 2015.
DOI: 10.1007/978-3-319-26784-5_13

In this work, we focus on developing PPR-estimators with *worst-case sublinear guarantees* for *undirected graphs*. Apart from their technical importance, our results are of practical relevance as several large-scale applications of PPR are based on undirected networks. For example, Facebook (which is an undirected social network) used Personalized PageRank for friend recommendation [5]. The social network Twitter is directed, but Twitter's friend recommendation algorithm (Who to Follow) [16] uses an algorithm called personalized SALSA [6,19], which first converts the directed network into an expanded undirected graph[1], and then computes PPR on this new graph. Random walks have also been used for collaborative filtering by the YouTube team [7] (on the undirected user-item bipartite graph), to predict future items a user will view. Applications like this motivate fast algorithms for PPR estimation on undirected graphs.

Equations (1) and (2) suggest two natural estimation algorithms for PPR – via linear-algebraic iterative techniques, and using Monte Carlo. The linear algebraic characterization of PageRank in Eq. (1) suggests the use of power iteration (or other localized iterations; cf Sect. 1.2 for details), while Eq. (2) is the basis for a Monte-Carlo algorithm, wherein we estimate $\pi_\sigma[t]$ by sampling independent L-step paths, each starting from a random state sampled from σ. For studying PageRank estimation algorithms, smaller probabilities are more difficult to estimate than large ones, so a natural parametrization is in terms of the minimum PageRank we want to detect. Formally, given any source σ, target node $t \in V$ and a desired minimum probability threshold δ, we want algorithms that give accurate estimates whenever $\pi_\sigma[t] \geq \delta$. Improved algorithms are motivated by the slow convergence of these algorithms: *both Monte Carlo and linear algebraic techniques have a running time of* $\Omega(1/\delta)$ *for PageRank estimation.* Furthermore this is true not only for worst case choices of target state t, but on average Monte-Carlo requires $\Omega(1/\delta)$ time to estimate a probability of size δ. Power iteration takes $\Theta(m)$ time, where m is the number of edges, and the work [21] shows empirically that the local version of power-iteration scales with $1/\delta$ for $\delta > 1/m$.

In a recent line of work, linear-algebraic and Monte-Carlo techniques were combined to develop new *bidirectional PageRank* estimators FAST-PPR [22] and Bidirectional-PPR [20], which gave the first significant improvement in the running-time of PageRank estimation since the development of Monte-Carlo techniques. Given an arbitrary source distribution σ and a *uniform random target node* t, these estimators were shown to return an accurate PageRank estimate with an *average* running-time of $\tilde{O}\left(\sqrt{\bar{d}/\delta}\right)$, where $\bar{d} = m/n$ is the average degree of the graph. Given $\tilde{O}\left(n\sqrt{\bar{d}/\delta}\right)$ precomputation and storage, the authors prove worst case guarantees for this bidirectional estimator but in practice that is a large precomputation requirement. This raised the challenge

---

[1] Specifically, for each node u in the original graph, SALSA creates two virtual nodes, a "consumer-node" u' and a "producer-node" u'', which are linked by an undirected edge. Any directed edge (u, v) is then converted into an undirected edge (u', v'') from u's consumer node to v's producer node.

of designing an algorithm with similar running-time guarantees over a *worst-case* choice of target node t. Inspired by the bidirectional estimators in [20,22], we propose a new PageRank estimator for *undirected graphs* with *worst-case* running time guarantees.

1.1   Our Contribution

We present the first estimator for personalized PageRank with sublinear running time in the worst case on undirected graphs. We formally present our `Undirected-BiPPR` algorithm in Sect. 2, and prove that it has the following accuracy and running-time guarantees:

Result 1 (See Theorem 1 in Sect. 2). *Given any undirected graph G, teleport probability α, source node s, target node t, threshold δ and relative error ϵ, the* `Undirected-BiPPR` *estimator (Algorithm 2) returns an unbiased estimate $\widehat{\pi}_s[t]$ for $\pi_s[t]$, which, with probability greater than $1 - p_{fail}$, satisfies:*

$$|\widehat{\pi}_s[t] - \pi_s[t]| < \max\{\epsilon\pi_s[t], 2e\delta\}.$$

Result 2 (See Theorem 2 in Sect. 2). *Let any undirected graph G, teleport probability α, threshold δ and desired relative error ϵ be given. For any source, target pair (s,t), the* `Undirected-BiPPR` *algorithm has a running-time of*

$$O\left(\frac{\sqrt{\ln(1/p_{fail})}}{\epsilon}\sqrt{\frac{d_t}{\delta}}\right), \text{ where } d_t \text{ is the degree of the target node } t.$$

In personalization applications, we are often only interested in personalized importance scores if they are greater than global importance scores, so it is natural to set δ based on the global importance of t. Assuming G is connected, in the limit $\alpha \to 0$, the PPR vector for any start node s converges to the stationary distribution of infinite-length random-walks on G – that is $\lim_{\alpha\to 0}\pi_s[t] = d_t/m$. This suggests that a natural PPR significance-test is to check whether $\pi_s(t) \geq d_t/m$. To this end, we have the following corollary:

Result 3 (See Corollary 1 in Sect. 2). *For any graph G and any (s,t) pair such that $\pi_s(t) \geq \frac{d_t}{m}$, then with high probability[2], *`Undirected-BiPPR`* returns an estimate $\pi_s(t)$ with relative error ϵ with a worst-case running-time of $O\left(\sqrt{m}\log n/\epsilon\right)$.*

Finally, in Sect. 3, using ideas from [8], we extend our technique to estimating more general random-walk transition-probabilities on undirected graphs, including graph diffusions and the heat kernel [11,18].

1.2   Existing Approaches for PageRank Estimation

We first summarize the existing methods for PageRank estimation:

Monte Carlo Methods: A standard method [4,9] for estimating $\pi_\sigma[t]$ is by using the terminal node of independently generated random walks of length

---

[2] Following convention, we use w.h.p. to mean with probability greater than $1 - \frac{1}{n}$.

$L \sim Geometric(\alpha)$ starting from a random node sampled from σ. Simple concentration arguments show that we need $\widetilde{\Theta}(1/\delta)$ samples to get an accurate estimate of $\pi_\sigma[t]$, irrespective of the choice of t and graph G.

Linear-Algebraic Iterations: Since the PageRank vector is the stationary distribution of a Markov chain, it can also be estimated via forward or reverse power iterations. A direct power iteration is often infeasible for large graphs; in such cases, it is preferable to use localized power iterations [1,2]. These local-update methods can also be used for other transition probability estimation problems such as heat kernel estimation [18]. Local update algorithms are often fast in practice, as unlike full power iteration methods they exploit the local structure of the chain. However even in sparse Markov chains and for a large fraction of target states, their running time can be $\Omega(1/\delta)$. For example, consider a random walk on a random d-regular graph and let $\delta = o(1/n)$. Then for $\ell \sim \log_d(1/\delta)$, verifying $\pi_{\mathbf{e}_s}[t] > \delta$ is equivalent to uncovering the entire $\log_d(1/\delta)$ neighborhood of s. However since a large random d-regular graph is (w.h.p.) an expander, this neighborhood has $\Omega(1/\delta)$ distinct nodes.

Bidirectional Techniques: Bidirectional methods are based on simultaneously working forward from the source node s and backward from the target node t in order to improve the running-time. One example of such a bidirectional technique is the use of *colliding random-walks* to estimate length-2ℓ random-walk transition probabilities in *regular undirected graphs* [14,17] – the main idea here is to exploit the reversibility by using two independent random walks of length ℓ starting from s and t respectively, and detecting if they collide. This results in reducing the number of walks required by a square-root factor, based on an argument similar to the birthday-paradox.

The `FAST-PPR` algorithm of Lofgren et al. [22] was the first bidirectional algorithm for estimating PPR in general graphs; this was subsequently refined and improved by the `Bidirectional-PPR` algorithm [20], and also generalized to other Markov chain estimation problems [8]. These algorithms are based on using a reverse local-update iteration from the target t (adapted from Andersen et al. [1]) to smear the mass over a larger *target set*, and then using random-walks from the source s to detect this target set. From a theoretical perspective, a significant breakthrough was in showing that for arbitrary choice of source node s these bidirectional algorithms achieved an *average* running-time of $\tilde{O}(\sqrt{d/\delta})$ over uniform-random choice of target node t – in contrast, both local-update and Monte Carlo has a running-time of $\Omega(1/\delta)$ for uniform-random targets. More recently, [10] showed that a similar bidirectional technique achieved a sublinear query-complexity for global PageRank computation, under a modified query model, in which all neighbors of a given node could be found in $O(1)$ time.

2   PageRank Estimation in Undirected Graphs

We now present our new bidirectional algorithm for PageRank estimation in undirected graphs.

2.1 Preliminaries

We consider an undirected graph $G(V, E)$, with n nodes and m edges. For ease of notation, we henceforth consider unweighted graphs, and focus on the simple case where $\sigma = \mathbf{e}_s$ for some single node s. We note however that all our results extend to weighted graphs and any source distribution σ in a straightforward manner.

2.2 A Symmetry for PPR in Undirected Graphs

The Undirected-BiPPR Algorithm critically depends on an underlying *reversibility property* exhibited by PPR vectors in undirected graphs. This property, stated before in several earlier works [3,15], is a direct consequence of the reversibility of random walks on undirected graphs. To keep our presentation self-contained, we present this property, along with a simple probabilistic proof, in the form of the following lemma:

Lemma 1. *Given any undirected graph G, for any teleport probability $\alpha \in (0,1)$ and for any node-pair $(s,t) \in V^2$, we have:*

$$\pi_s[t] = \frac{d_t}{d_s}\pi_t[s].$$

Proof. For path $P = \{s, v_1, v_2, \ldots, v_k, t\}$ in G, we denote its length as $\ell(P)$ (here $\ell(P) = k + 1$), and define its reverse path to be $\overline{P} = \{t, v_k, \ldots, v_2, v_1, s\}$ – note that $\ell(P) = \ell(\overline{P})$. Moreover, we know that a random-walk starting from s traverses path P with probability $\mathbb{P}[P] = \frac{1}{d_s} \cdot \frac{1}{d_{v_1}} \cdot \ldots \cdot \frac{1}{d_{v_k}}$, and thus, it is easy to see that we have:

$$\mathbb{P}[P] \cdot d_s = \mathbb{P}[\overline{P}] \cdot d_t \tag{3}$$

Now let $\mathcal{P}_{st}$ denote the set of paths in G starting at s and terminating at t. Then we can re-write Eq. (2) as:

$$\pi_s[t] = \prod_{P \in \mathcal{P}_{st}} \alpha(1-\alpha)^{\ell(P)}\mathbb{P}[P] = \prod_{\overline{P} \in \mathcal{P}_{ts}} \alpha(1-\alpha)^{\ell(\overline{P})}\mathbb{P}[\overline{P}] = \frac{d_t}{d_s}\pi_t[s] \qquad \square$$

2.3 The Undirected-BiPPR Algorithm

At a high level, the Undirected-BiPPR algorithm has two components:

– **Forward-work:** Starting from source s, we first use a forward local-update algorithm, the ApproximatePageRank$(G, \alpha, s, r_{\max})$ algorithm of Andersen et al. [2] (shown here as Algorithm 1). This procedure begins by placing one unit of "residual" probability-mass on s, then repeatedly selecting some node u, converting an α-fraction of the residual mass at u into probability mass, and pushing the remaining residual mass to u's neighbors. For any node u, it returns an estimate $p_s[u]$ of its PPR $\pi_s[u]$ from s as well as a residual $r_s[u]$ which represents un-pushed mass at u.

– **Reverse-work:** We next sample random walks of length $L \sim Geometric(\alpha)$ starting from t, and use the residual at the terminal nodes of these walks to compute our desired PPR estimate. Our use of random walks backwards from t depends critically on the symmetry in undirected graphs presented in Lemma 1.

Note that this is in contrast to `FAST-PPR` and `Bidirectional-PPR`, which performs the local-update step in reverse from the target t, and generates random-walks forwards from the source s.

Algorithm 1. `ApproximatePageRank`$(G, \alpha, s, r_{\max})$ [2]

Inputs: graph G, teleport probability α, start node s, maximum residual $r_{\max}$
1: Initialize (sparse) estimate-vector $p_s = \mathbf{0}$ and (sparse) residual-vector $r_s = e_s$
   (i.e. $r_s[v] = 1$ if $v = s$; else 0)
2: **while** $\exists u \in V$ s.t. $\frac{r_s[u]}{d_u} > r_{\max}$ **do**
3:   **for** $v \in \mathcal{N}[u]$ **do**
4:      $r_s[v] += (1 - \alpha)r_s[u]/d_u$
5:   **end for**
6:   $p_s[u] += \alpha r_s[u]$
7:   $r_s[u] = 0$
8: **end while**
9: **return** (p_s, r_s)

In more detail, our algorithm will choose a maximum residual parameter $r_{\max}$, and apply the local push operation in Algorithm 1 until for all v, $r_s[v]/d_v < r_{\max}$. Andersen et al. [2] prove that their local-push operation preserves the following invariant for vectors (p_s, r_s):

$$\pi_s[t] = p_s[t] + \sum_{v \in V} r_s[v]\pi_v[t], \qquad \forall t \in V. \tag{4}$$

Since we ensure that $\forall v, r_s[v]/d_v < r_{\max}$, it is natural at this point to use the symmetry Lemma 1 and re-write this as:

$$\pi_s[t] = p_s[t] + d_t \sum_{v \in V} \frac{r_s[v]}{d_v}\pi_t[v].$$

Now using the fact that $\sum_t \pi_v[t] = n\pi[t]$ get that $\forall t \in V$, $|\pi_s[t] - p_s[t]| \leq r_{\max}d_t n\pi[t]$.

However, we can get a more accurate estimate by using the residuals. The key idea of our algorithm is to re-interpret this as an expectation:

$$\pi_s[t] = p_s[t] + d_t \mathbb{E}_{V \sim \pi_t}\left[\frac{r_s[v]}{d_V}\right]. \tag{5}$$

We estimate the expectation using standard Monte-Carlo. Let $V_i \sim \pi_t$ and $X_i = r_s(V_i)d_t/d_{V_i}$, so we have $\pi_s[t] = p_s[t] + \mathbb{E}[X]$. Moreover, each sample X_i is bounded by $d_t r_{max}$ (this is the stopping condition for ApproximatePageRank), which allows us to efficiently estimate its expectation. To this end, we generate w random walks, where

$$w = \frac{c}{\epsilon^2} \frac{r_{max}}{\delta/d_t}.$$

The choice of c is specified in Theorem 1. Finally, we return the estimate:

$$\widehat{\pi}_s[t] = p_t[s] + \frac{1}{w} \sum_{i=1}^{w} X_i.$$

The complete pseudocode is given in Algorithm 2.

Algorithm 2. Undirected-BiPPR(s, t, δ)

Inputs: graph G, teleport probability α, start node s, target node t, minimum probability δ, accuracy parameter $c = 3\ln(2/p_{fail})$ (cf. Theorem 1)

1: $(p_s, r_s) = $ ApproximatePageRank(s, r_{max})
2: Set number of walks $w = cd_t r_{max}/(\epsilon^2 \delta)$
3: **for** index $i \in [w]$ **do**
4:    Sample a random walk starting from t, stopping after each step with probability α; let V_i be the endpoint
5:    Set $X_i = r_s(V_i)/d_{V_i}$
6: **end for**
7: **return** $\widehat{\pi}_s[t] = p_s[t] + (1/w)\sum_{i\in[w]} X_i$

2.4   Analyzing the Performance of Undirected-BiPPR

Accuracy Analysis: We first prove that Undirected-BiPPR returns an unbiased estimate with the desired accuracy:

Theorem 1. *In an undirected graph G, for any source node s, minimum threshold δ, maximum residual r_{max}, relative error ϵ, and failure probability p_{fail}, Algorithm 2 outputs an estimate $\widehat{\pi}_s[t]$ such that with probability at least $1 - p_{fail}$ we have: $|\pi_s[t] - \hat{\pi}_s[t]| \leq \max\{\epsilon\pi_s[t], 2e\delta\}$.*

The proof follows a similar outline as the proof of Theorem 1 in [20]. For completeness, we sketch the proof here:

Proof. As stated in Algorithm 2, we average over $w = cd_t r_{max}/\epsilon^2\delta$ walks, where c is a parameter we choose later. Each walk is of length $Geometric(\alpha)$, and we denote V_i as the last node visited by the ith walk; note that $V_i \sim \pi_t$. As defined above, let $X_i = r_s(V_i)d_t/d_{V_i}$; the estimate returned by Undirected-BiPPR is:

$$\widehat{\pi}_s[t] = p_t[s] + \frac{1}{w} \sum_{i=1}^{w} X_i.$$

First, from Eq. (5), we have that $\mathbb{E}[\widehat{\pi}_s[t]] = \pi_s[t]$. Also, `ApproximatePageRank` guarantees that for all v, $r_s[v] < d_v r_{\max}$, and so each X_i is bounded in $[0, d_t r_{\max}]$; for convenience, we rescale X_i by defining $Y_i = \frac{1}{d_t r_{\max}} X_i$.

We now show concentration of the estimates via the following Chernoff bounds (see Theorem 1.1 in [12]):

1. $\mathbb{P}[|Y - \mathbb{E}[Y]| > \epsilon \mathbb{E}[Y]] < 2 \exp(-\frac{\epsilon^2}{3} \mathbb{E}[Y])$
2. For any $b > 2e\mathbb{E}[Y]$, $\mathbb{P}[Y > b] \le 2^{-b}$

We perform a case analysis based on whether $\mathbb{E}[X_i] \ge \delta$ or $\mathbb{E}[X_i] < \delta$. First, if $\mathbb{E}[X_i] \ge \delta$, then we have $\mathbb{E}[Y] = \frac{w}{d_t r_{\max}} \mathbb{E}[X_i] = \frac{c}{\epsilon^2 \delta} \mathbb{E}[X_i] \ge \frac{c}{\epsilon^2}$, and thus:

$$\mathbb{P}\left[|\widehat{\pi}_s[t] - \pi_s[t]| > \epsilon \pi_s[t]\right] \le \mathbb{P}\left[|\bar{X} - \mathbb{E}[X_i]| > \epsilon \mathbb{E}[X_i]\right] = \mathbb{P}\left[|Y - \mathbb{E}[Y]| > \epsilon \mathbb{E}[Y]\right]$$

$$\le 2 \exp\left(-\frac{\epsilon^2}{3} \mathbb{E}[Y]\right) \le 2 \exp\left(-\frac{c}{3}\right) \le p_{\text{fail}},$$

where the last line holds as long as we choose $c \ge 3 \ln(2/p_{\text{fail}})$.

Suppose alternatively that $\mathbb{E}[X_i] < \delta$. Then:

$$\mathbb{P}[|\widehat{\pi}_s[t] - \pi_s[t]| > 2e\delta] = \mathbb{P}[|\bar{X} - \mathbb{E}[X_i]| > 2e\delta] = \mathbb{P}\left[|Y - \mathbb{E}[Y]| > \frac{w}{d_t r_{\max}} 2e\delta\right]$$

$$\le \mathbb{P}\left[Y > \frac{w}{d_t r_{\max}} 2e\delta\right].$$

At this point we set $b = 2e\delta w/d_t r_{\max} = 2ec/\epsilon^2$ and apply the second Chernoff bound. Note that $\mathbb{E}[Y] = c\mathbb{E}[X_i]/\epsilon^2\delta < c/\epsilon^2$, and hence we satisfy $b > 2e\mathbb{E}[Y]$. We conclude that:

$$\mathbb{P}[|\widehat{\pi}_s[t] - \pi_s[t]| > 2e\delta] \le 2^{-b} \le p_{\text{fail}}$$

as long as we choose c such that $c \ge \frac{\epsilon^2}{2e} \log_2 \frac{1}{p_{\text{fail}}}$. The proof is completed by combining both cases and choosing $c = 3 \ln(2/p_{\text{fail}})$. □

Running Time Analysis: The more interesting analysis is that of the running-time of `Undirected-BiPPR` – we now prove a worst-case running-time bound:

Theorem 2. *In an undirected graph, for any source node (or distribution) s, target t with degree d_t, threshold δ, maximum residual r_{max}, relative error ϵ, and failure probability p_{fail}, Undirected-BiPPR has a worst-case running-time of:*

$$O\left(\frac{\sqrt{\log \frac{1}{p_{fail}}}}{\epsilon} \sqrt{\frac{d_t}{\delta}}\right).$$

Before proving this result, we first state and prove a crucial lemma from [2]:

Lemma 2 (Lemma 2 in [2]). *Let T be the total number of push operations per-formed by ApproximatePageRank, and let d_k be the degree of the vertex involved in the kth push. Then:*

$$\sum_{k=1}^{T} d_k \leq \frac{1}{\alpha r_{max}}$$

Proof. Let v_k be the vertex pushed in the kth step – then by definition, we have that $r_s(v_k) > r_{max} d_k$. Now after the local-push operation, the sum residual $||r_s||_1$ decreases by at least $\alpha r_{max} d_k$. However, we started with $||r_s||_1 = 1$, and thus we have $\sum_{k=1}^{T} \alpha r_{max} d_k \leq 1$. □

Note also that the amount of work done while pushing from a node v is d_v.

Proof (of Theorem 2). As proven in Lemma 2, the push forward step takes total time $O(1/\alpha r_{max})$ in the *worst-case*. The random walks take $O(w) = O\left(\frac{1}{\epsilon^2} \frac{r_{max}}{\delta/d_t}\right)$ time. Thus our total time is

$$O\left(\frac{1}{r_{max}} + \frac{\ln \frac{1}{p_{fail}}}{\epsilon^2} \frac{r_{max}}{\delta/d_t}\right).$$

Balancing this by choosing $r_{max} = \frac{\epsilon}{\sqrt{\ln \frac{1}{p_{fail}}}} \sqrt{\delta/d_t}$, we get total running-time:

$$O\left(\frac{\sqrt{\ln \frac{1}{p_{fail}}}}{\epsilon} \sqrt{\frac{d_t}{\delta}}\right). \qquad \square$$

We can get a cleaner worst-case running time bound if we make a natural assumption on $\pi_s[t]$. In an undirected graph, if we let $\alpha = 0$ and take infinitely long walks, the stationary probability of being at any node t is $\frac{d_t}{m}$. Thus if $\pi_s[t] < \frac{d_t}{m}$, then s actually has a lower PPR to t than the non-personalized stationary probability of t, so it is natural to say t is not significant for s. If we set a significance threshold of $\delta = \frac{d_t}{m}$, and apply the previous theorem, we immediately get the following:

Corollary 1. *If $\pi_s[t] \geq \frac{d_t}{m}$, we can estimate $\pi_s[t]$ within relative error ϵ with probability greater than $1 - \frac{1}{n}$ in worst-case time:*

$$O\left(\frac{\log n}{\epsilon} \sqrt{m}\right).$$

In contrast, the running time for Monte-Carlo to achieve the same accuracy guarantee is $O\left(\frac{1}{\delta} \frac{\log(1/p_{fail})}{\alpha \epsilon^2}\right)$, and the running time for ApproximatePageRank is $O\left(\frac{d}{\delta \alpha}\right)$. The FAST-PPR algorithm of [22] has an *average case* running time of

$$O\left(\frac{1}{\alpha \epsilon^2} \sqrt{\frac{d}{\delta}} \sqrt{\frac{\log(1/p_{fail}) \log(1/\delta)}{\log(1/(1-\alpha))}}\right)$$ for uniformly chosen targets, but has no clean worst-case running time bound because its running time depends on the degree of nodes pushed from in the linear-algebraic part of the algorithm.

3   Extension to Graph Diffusions

PageRank and Personalized PageRank are a special case of a more general set of network-centrality metrics referred to as *graph diffusions* [11,18]. In a graph diffusion we assign a weight α_i to walks of length i. The score is then is a polynomial function of the random-walk transition probabilities of the form:

$$f(W, \sigma) := \sum_{i=0}^{\infty} \alpha_i \left(\sigma W^i \right),$$

where $\alpha_i \geq 0, \sum_i \alpha_i = 1$. To see that PageRank has this form, we can expand Eq. (1) via a Taylor series to get:

$$\pi_\sigma = \sum_{i=1}^{\infty} \alpha(1 - \alpha)^i \left(\sigma W^i \right)$$

Another important graph diffusion is the *heat kernel* h_σ, which corresponds to the scaled matrix exponent of $(I - W)^{-1}$:

$$h_{\sigma,\gamma} = e^{-\gamma(I-W)^{-1}} = \sum_{i=1}^{\infty} \frac{e^{-\gamma}\gamma^i}{i!} \left(\sigma W^i \right)$$

In [8], Banerjee and Lofgren extended `Bidirectional-PPR` to get bidirectional estimators for graph diffusions and other general Markov chain transition-probability estimation problems. These algorithms inherited similar performance guarantees to `Bidirectional-PPR` – in particular, they had good expected running-time bounds for uniform-random choice of target node t. We now briefly discuss how we can modify `Undirected-BiPPR` to get an estimator for graph diffusions in undirected graphs with worst-case running-time bounds.

First, we observe that Lemma 1 extends to all graph diffusions, as follows:

Corollary 2. *Let any undirected graph G with random-walk matrix W, and any set of non-negative length weights $(\alpha_i)_{i=0}^{\infty}$ with $\sum \alpha_i = 1$ be given. Define $f(W, \sigma) = \sum_{i=0}^{\infty} \alpha_i \left(\sigma W^i \right)$. Then for any node-pair $(s,t) \in V^2$, we have:*

$$f\left(W, \underline{e}_s\right) = \frac{d_t}{d_s} f\left(W, \underline{e}_t\right).$$

As before, the above result is stated for unweighted graphs, but it also extends to random-walks on weighted undirected graphs, if we define $d_i = \sum_j w_{ij}$.

Next, observe that for any graph diffusion $f(\cdot)$, the truncated sum $f^{\ell_{max}} = \sum_{i=0}^{\ell_{max}} \alpha_i \left(\pi_\sigma^T P^i \right)$ obeys: $||f - f^{\ell_{max}}||_\infty \leq \sum_{\ell_{max}+1}^{\infty} \alpha_k$. Thus a guarantee on an estimate for the truncated sum directly translates to a guarantee on the estimate for the diffusion.

The main idea in [8] is to generalize the bidirectional estimators for PageRank to estimating *multi-step transitions probabilities* (for short, MSTP). Given a source node s, a target node t, and length $\ell \leq \ell_{max}$, we define:

$$p_s^\ell[t] = \mathbb{P}[\text{Random-walk of length } \ell \text{ starting from } s \text{ terminates at } t]$$

Note from Corollary 2, we have for any pair (s,t) and any ℓ, $p_s^\ell[t]d_s = p_t^\ell[s]d_t$.

Now in order to develop a bidirectional estimator for $p_s^\ell[t]$, we need to define a local-update step similar to `ApproximatePageRank`. For this, we can modify the `REVERSE-PUSH` algorithm from [8], as follows.

Similar to `ApproximatePageRank`, given a source node s and maximum length ℓ_{max}, we associate with each length $\ell \leq \ell_{max}$ an estimate vector q_s^ℓ and a residual vector r_s^ℓ. These are updated via the following `ApproximateMSTP` algorithm:

Algorithm 3. ApproximateMSTP$(G, s, \ell_{max}, r_{max})$

Inputs: Graph G, source s, maximum steps ℓ_{max}, maximum residual r_{max}

1: Initialize: Estimate-vectors $q_s^k = \underline{0}$, $\forall k \in \{0, 1, 2, \ldots, \ell_{max}\}$,
       Residual-vectors $r_s^0 = \underline{e}_s$ and $r_s^k = \underline{0}$, $\forall k \in \{1, 2, 3, \ldots, \ell_{max}\}$

2: **for** $i \in \{0, 1, \ldots, \ell_{max}\}$ **do**

3:    **while** $\exists v \in \S$ *s.t.* $r_t^i[v]/d_v > r_{max}$ **do**

4:       **for** $w \in \mathcal{N}(v)$ **do**

5:          $r_s^{i+1}[w] \mathrel{+}= r_s^i[v]/d_v$

6:       **end for**

7:       $q_s^i[v] \mathrel{+}= r_s^i[v]$

8:       $r_s^i[v] = 0$

9:    **end while**

10: **end for**

11: **return** $\{q_s^\ell, r_s^\ell\}_{\ell=0}^{\ell_{max}}$

The main observation now is that for any source s, target t, and length ℓ, after executing the `ApproximateMSTP` algorithm, the vectors $\{q_s^\ell, r_s^\ell\}_{\ell=0}^{\ell_{max}}$ satisfy the following invariant (via a similar argument as in [8], Lemma 1):

$$p_s^\ell[t] = q_s^\ell[t] + \sum_{k=0}^{\ell} \sum_{v \in V} r_s^k[v] p_v^{\ell-k}[t] = q_s^\ell[t] + d_t \sum_{k=0}^{\ell} \sum_{v \in V} \frac{r_s^k[v]}{d_v} p_t^{\ell-k}[v]$$

As before, note now that the last term can be written as an expectation over random-walks originating from t. The remaining algorithm, accuracy analysis, and runtime analysis follow the same lines as those in Sect. 2.

4   Conclusion

We have developed `Undirected-BiPPR`, a new bidirectional PPR-estimator for undirected graphs, which for any (s,t) pair such that $\pi_s[t] > d_t/m$, returns an estimate with ϵ relative-error in worst-case running time of $O(\sqrt{m}/\epsilon)$. This thus extends the average-case running-time improvements achieved in [20,22] to worst-case bounds on undirected graphs, using the reversibility of random-walks on undirected graphs. Whether such worst-case running-time results extend to general graphs, or if PageRank computation is fundamentally easier on undirected graphs as opposed to directed graphs, remains an open question.

Acknowledgments. Research supported by the DARPA GRAPHS program via grant FA9550-12-1-0411, and by NSF grant 1447697. One author was supported by an NPSC fellowship. Thanks to Aaron Sidford for a helpful discussion.

References

1. Andersen, R., Borgs, C., Chayes, J., Hopcraft, J., Mirrokni, V.S., Teng, S.-H.: Local computation of pagerank contributions. In: Bonato, A., Chung, F.R.K. (eds.) WAW 2007. LNCS, vol. 4863, pp. 150–165. Springer, Heidelberg (2007)
2. Andersen, R., Chung, F., Lang, K.: Local graph partitioning using pagerank vectors. In: 47th Annual IEEE Symposium on Foundations of Computer Science, FOCS 2006 (2006)
3. Avrachenkov, K., Gonçalves, P., Sokol, M.: On the choice of kernel and labelled data in semi-supervised learning methods. In: Bonato, A., Mitzenmacher, M., Prałat, P. (eds.) WAW 2013. LNCS, vol. 8305, pp. 56–67. Springer, Heidelberg (2013)
4. Avrachenkov, K., Litvak, N., Nemirovsky, D., Osipova, N.: Monte carlo methods in pagerank computation: when one iteration is sufficient. SIAM J. Numer. Anal. **45**, 890–904 (2007)
5. Backstrom, L., Leskovec, J.: Supervised random walks: predicting and recommending links in social networks. In: Proceedings of the Fourth ACM International Conference on Web Search and Data Mining. ACM (2011)
6. Bahmani, B., Chowdhury, A., Goel, A.: Fast incremental and personalized pagerank. Proc. VLDB Endowment **4**(3), 173–184 (2010)
7. Baluja, S., Seth, R., Sivakumar, D., Jing, Y., Yagnik, J., Kumar, S., Ravichandran, D., Aly, M.: Video suggestion and discovery for youtube: taking random walks through the view graph. In: Proceedings of the 17th International Conference on World Wide Web. ACM (2008)
8. Banerjee, S., Lofgren, P.: Fast bidirectional probability estimation in markov models. In: NIPS (2015)
9. Borgs, C., Brautbar, M., Chayes, J., Teng, S.-H.: A sublinear time algorithm for pagerank computations. In: Bonato, A., Janssen, J. (eds.) WAW 2012. LNCS, vol. 7323, pp. 41–53. Springer, Heidelberg (2012)
10. Bressan, M., Peserico, E., Pretto, L.: Approximating pagerank locally with sublinear query complexity. arXiv preprint arXiv:1404.1864 (2014)
11. Chung, F.: The heat kernel as the pagerank of a graph. Proc. Nat. Acad. Sci. **104**, 19735–19740 (2007)
12. Dubhashi, D., Panconesi, A.: Concentration of Measure for the Analysis of Randomized Algorithms. Cambridge University Press, New York (2009)
13. Gleich, D.F.: PageRank beyond the web. arXiv, cs.SI:1407.5107 (2014). Accepted for publication in SIAM Review

14. Goldreich, O., Ron, D.: On testing expansion in bounded-degree graphs. In: Goldreich, O. (ed.) Studies in Complexity and Cryptography. LNCS, vol. 6650, pp. 68–75. Springer, Heidelberg (2011)

15. Grolmusz, V.: A note on the pagerank of undirected graphs. Inf. Process. Lett. **115**, 633–634 (2015)

16. Gupta, P., Goel, A., Lin, J., Sharma, A., Wang, D., Zadeh, R.: Wtf: the who to follow service at twitter. In: Proceedings of the 22nd International Conference on World Wide Web, pp. 505–514. International World Wide Web Conferences Steering Committee (2013)

17. Kale, S., Peres, Y., Seshadhri, C.: Noise tolerance of expanders and sublinear expander reconstruction. In: Proceedings of the IEEE FOCS 2008. IEEE (2008)

18. Kloster, K., Gleich, D.F.: Heat kernel based community detection. In: Proceedings of the ACM SIGKDD 2014 (2014)

19. Lempel, R., Moran, S.: The stochastic approach for link-structure analysis (salsa) and the tkc effect. Comput. Netw. **33**(1), 387–401 (2000)

20. Lofgren, P., Banerjee, S., Goel, A.: Personalized pagerank estimation and search: A bidirectional approach. Technical report (2015)

21. Lofgren, P., Goel, A.: Personalized pagerank to a target node. arXiv preprint arXiv:1304.4658 (2013)

22. Lofgren, P.A., Banerjee, S., Goel, A., Seshadhri, C.: FAST-PPR: scaling personalized pagerank estimation for large graphs. In: Proceedings of the ACM SIGKDD 2014. ACM (2014)

23. Page, L., Brin, S., Motwani, R., Winograd, T.: The pagerank citation ranking: bringing order to the web (1999)

Distributed Algorithms for Finding Local Clusters Using Heat Kernel Pagerank

Fan Chung and Olivia Simpson(✉)

Department of Computer Science and Engineering, University of California,
San Diego, La Jolla, CA 92093, USA
{fan,osimpson}@ucsd.edu

Abstract. We consider the problem of computing local clusters in large graphs distributed across nodes in a network using two different models of distributed computation. We give a distributed algorithm that computes a local cluster in time that depends only logarithmically on the size of the graph in the CONGEST model. In particular, when the conductance of the optimal local cluster is known, the algorithm runs in time entirely independent of the size of the graph and depends only on error bounds for approximation. We also show that the local cluster problem can be computed in the k-machine distributed model in sublinear time. The speedup of our local cluster algorithms is mainly due to the use of our distributed algorithm for heat kernel pagerank.

Keywords: Distributed algorithms · Local cluster · Sparse cut · Heat kernel pagerank · Heat kernel · Random walk

1 Introduction

Distributed computation is an increasingly important framework as the demand for fast data analysis grows and data simultaneously becomes too large to fit in main memory. As distributed systems for large-scale graph processing such as Pregel [21], GraphLab [20], and Google's MapReduce [12] are rapidly developing, there is a need for both theoretical and practical bounds in adapting classical graph algorithms to a modern distributed and parallel setting.

A distributed algorithm performs local computations on pieces of input and communicates the results through given communication links. When processing a massive graph in a distributed algorithm, local outputs must be configured without shared memory and with few rounds of communication. A central problem of interest is to compute local clusters in large graphs in a distributed setting.

Computing local clusters are of certain application-specific interests, such as detecting communities in social networks [16] or groups of interacting proteins in biological networks [17]. When the graph models the computer network itself, detecting local clusters can help identify communication bottlenecks, where one set of well-connected nodes is separated from another by a small number of links. Further, being able to identify the clusters quickly prevents bottlenecks from developing as the network grows.

© Springer International Publishing Switzerland 2015
D.F. Gleich et al. (Eds.): WAW 2015, LNCS 9479, pp. 177–189, 2015.
DOI: 10.1007/978-3-319-26784-5_14

A local clustering algorithm computes a set of vertices in a graph with a small Cheeger ratio (or so-called conductance as defined in Sect. 2.2). Moreover, we ask that the algorithm use only local information. In the static setting, an important consequence of this locality constraint is running times proportional to the size of the output set, rather than the entire graph. In this paper, we present the first algorithms for computing local clusters in two distributed settings that finish in a sublinear number of rounds of communication.

A standard technique in local clustering algorithms is the so-called *sweep* algorithm. In a sweep, one orders the vertices of a graph according to some real-valued function defined on the vertex set and then investigates the cut set induced by each prefix of vertices in the ordering. The classical method of spectral clustering uses eigenvectors as functions for the sweep. For local clustering algorithms, the sweep functions are based on random walks [18,19,25]. In [1], the efficiency of the local clustering algorithm is due to the use of PageRank vectors as the sweep functions [4]. In this paper, the main leverage in the improved running times of our algorithms is to use the heat kernel pagerank vector for performing a sweep. In particular, we are able to exploit parallelism in our algorithm for computing the heat kernel pagerank and give a distributed random walk-based procedure which requires fewer rounds of communication and yet maintains similar approximation guarantees as previous algorithms.

In Sect. 2.1, we will describe two distributive models – the CONGEST model and the k-machine model. We demonstrate in two different distributed settings that a heat kernel pagerank distribution can be used to compute local clusters with Cheeger ratio $O(\sqrt{\phi})$ when the optimal local cluster has Cheeger ratio ϕ. With a fast, parallel algorithm for approximating the heat kernel pagerank and efficient local computations, our algorithm works on an n-vertex graph in the CONGEST, or standard message passing, model with high probability in at most $O\left(\frac{\log(\epsilon^{-1})\log n}{\log\log(\epsilon^{-1})} + \frac{1}{\epsilon}\log n\right)$ rounds of communication where ϵ is an error bound for approximation. This is an improvement over the previously best-performing local clustering algorithm in [9] which uses a personalized PageRank vector and finishes in $O\left(\frac{1}{\alpha}\log^2 n + n\log n\right)$ rounds in the CONGEST model for any $0 < \alpha < 1$. We then extend our results to the k-machine model to show that a local cluster can be computed in $\tilde{O}\left(\frac{\log(\epsilon^{-1})}{\epsilon^3 k^2 \log\log(\epsilon^{-1})} + \frac{1}{\epsilon k^2} + \left(\frac{\log(\epsilon^{-1})}{k\log\log(\epsilon^{-1})} + \frac{1}{k\epsilon}\right)\max\left\{\frac{1}{\epsilon^3}, \Delta\right\}\right)$ rounds, where Δ is the maximum degree in the graph, with high probability.

1.1   Related Work

The idea of computing local clusters with random walks was introduced by Lovász and Simonovits in their works analyzing the isoperimetric properties of random walks on graphs [18,19]. Spielman and Teng [25] expanded upon these ideas and gave the first nearly-linear time algorithm for local clustering, improving the original framework by sparsifying the graph. The algorithm of [25] finds a local cluster with Cheeger ratio $O(\sqrt{\phi}\log^{3/2} n)$ in time $O(m(\log n/\phi)^{O(1)})$, where m is the number of edges in the graph. Each of these algorithms uses the distribution of random walks of length $O(\frac{1}{\phi})$. Andersen et al. [1] give a local clustering algorithm using the distribution given by a PageRank vector. Their algorithm promises a $O(\sqrt{\phi}\log^{1/2}n)$ cluster approximation and runs in time $O(\frac{m}{\phi}\log^4 m)$. Orecchia et al. use a variant of heat kernel random walks in their randomized algorithm for computing a cut in a graph with prescribed balance constraints [22]. A key subroutine in the algorithm is a procedure for computing $e^{-A}v$ for a positive semidefinite matrix A and a unit vector v in time $\tilde{O}(m)$ for graphs on n vertices and m edges. Indeed, heat kernel has proven to be an efficient and effective tool for local cluster detection [6,15].

Andersen and Peres [2] simulate a volume-biased evolving set process to find sparse cuts. Their algorithm improves the ratio between the running time of the algorithm on a given run and the volume of the output set while maintaining similar approximation guarantees as previous algorithms. Their algorithm is later improved in [13]. Arora, Rao, and Vazirani [3] give a $O(\sqrt{\log n})$-approximation algorithm using semi-definite programming techniques, however it is slower than algorithms based on spectral methods and random walks.

For distributed algorithms, in [11] fast random walk-based distributed algorithms are given for estimating mixing time, conductance and the spectral gap of a network. In [10], distributed algorithms are derived for computing PageRank vectors with $O(\frac{1}{\alpha}\log n)$ rounds for any $0 < \alpha < 1$ with high probability. Das Sarma et al. [9] give two algorithms for computing sparse cuts in the CONGEST distributed model. The first algorithm uses random walks and is based on the analysis of [25]. By incorporating the results of [11], they show that the stationary distribution of a random walk of length l can be computed in $O(l)$ rounds. The second algorithm in [9] uses PageRank vectors and is based on the analysis of [1]. By using the results of [10], the authors of [9] compute local clusters in $O((\frac{1}{\phi}+n)\log n)$ rounds with standard random walks and $O(\frac{1}{\alpha}\log^2 n + n\log n)$ rounds using PageRank vectors.

2   The Setting and Our Contributions

2.1   Models of Computation

We consider two models of distributed computation – the CONGEST model and the k-machine model. In each, data is distributed across nodes (machines) of a network which may communicate over specified communication links in rounds. Memory is decentralized, and the goal is to minimize the running time by minimizing the number of rounds required for computation for an arbitrary input graph G. We emphasize that local communication is taken to be free.

The CONGEST Model. The first model we consider is the CONGEST model. In this model, the communication links are exactly the edges of the input graph and each vertex is mapped to a dedicated machine. The CONGEST (or standard message-passing) model was introduced in [23,24] to simulate real-world bandwidth restrictions across a network.

Due to how the vertices are distributed in the network, we simplify the model by assuming the computer network is the input graph $G = (V, E)$ on $n = |V|$ nodes or machines and $m = |E|$ edges or communication links. Each node has a unique $\log n$-bit ID. Initially each node only possesses its own ID and the IDs of each of its neighbors, and in some instances we may allow nodes some metadata about the graph (the value of n, for instance). Nodes can only communicate through edges of the network and communication occurs in rounds. That is, any message sent at the beginning of round r is fully transmitted and received by the end of round r. We assume that all nodes run with the same processing speed. Most importantly, we only allow $O(\log n)$ bits to be transmitted across any edge per round.

The k-machine Model. The defining difference between the k-machine model and the CONGEST model is that, whereas vertices are mapped to distinct, dedicated machines in the CONGEST model, a number of vertices may be mapped to the same machine in the k-machine model. This model is meant to more accurately simulate distributed graph computation in systems such as Pregel [21] and GraphLab [20].

We consider computing over massive datasets distributed over nodes of the k-machine network. The complete data is never known by any individual machine, and there is no shared memory. Each machine executes an instance of a distributed algorithm, and the output of each machine is with respect to the data points it hosts. A solution to a full problem is then a particular configuration of the outputs of each of the machines. The model is discussed in greater detail in Sect. 5.

The two models are limiting and advantageous in different ways, and one is not inherently better than the other. For instance, since many vertices are mapped to a single machine in the k-machine model, there is more "local information" available since vertices sharing a machine can communicate for free. However, since communication is restricted to the communication links in the computer network, vertex-vertex communication is somewhat less restrictive in the CONGEST model since links exactly correspond to edges. The consequences of these differences are largely observed in time complexity, and certain graph problems are more suited to one model than the other.

In this paper we analyze our algorithmic techniques in the CONGEST model, and then use the Conversion Theorem of [14] to give an efficient probabilistic algorithm in the k-machine model for computing local clusters.

2.2 Local Clusters and Heat Kernel Pagerank

Throughout this paper, we consider a graph $G = (V, E)$ with $n = |V|$ and $m = |E|$ that is connected and undirected. In this section we give some definitions that will make our problem statement and results precise.

Personalized Heat Kernel Pagerank. The heat kernel pagerank is so named for the *heat kernel* of the graph, $\mathcal{H}_t = e^{-t\mathcal{L}}$, where $\mathcal{L}$ is the normalized graph Laplacian $\mathcal{L} = D^{-1/2}(D-A)D^{-1/2}$. Here D is the diagonal matrix whose entries correspond to vertex degree and A is the symmetric adjacency matrix. The heat kernel is a solution to the heat equation $\frac{\partial u}{\partial t} = -\mathcal{L}u$, and thus has fundamental connections to diffusion properties of a graph. Because of its connection to random walks, for heat kernel pagerank we use a similar heat kernel matrix, $H_t = e^{-t\mathbf{L}}$, where $\mathbf{L} = I - P$. Here, I is the $n \times n$ identity matrix and $P = D^{-1}A$ is the transition probability matrix corresponding to the following standard random walk on the graph: at each step, move from a vertex v to a random neighbor u. Then the *heat kernel pagerank* is defined in terms of a preference (row) vector f as $\rho_{t,f} = fH_t$. When f, as a row vector, is some probability distribution over the vertices, the following formulation is useful for our Monte Carlo-based approximation algorithm:

$$\rho_{t,f} = fH_t = \sum_{k=0}^{\infty} e^{-t}\frac{t^k}{k!}fP^k. \tag{1}$$

In this paper, we consider preference vectors $f = \chi_s$ with all probability on a single vertex s, called the *seed*, and zero probability elsewhere. This is a common starting distribution for the PageRank vector, as well, commonly referred to as a *personalized PageRank* (or PPR) vector. We will adapt similar terminology and refer to the vector $\rho_{t,s} := \rho_{t,\chi_s}$ as the *personalized heat kernel pagerank vector for s*, or simply PHKPR.

Cheeger Ratio. For a non-empty subset $S \subset V$ of vertices in a graph, define the *volume* to be $\mathrm{vol}(S) = \sum_{v \in S} d_v$, where d_v is the degree of vertex v. The *Cheeger ratio* of a set S is defined as $\varPhi(S) = \frac{|E(S,\bar{S})|}{\min\{\mathrm{vol}(S),\mathrm{vol}(\bar{S})\}}$, where we use $\bar{S}$ here to denote the set $V\backslash S$, and $E(S,\bar{S})$ is the set of edges with one endpoint in S and the other in $\bar{S}$. The Cheeger ratio of a graph, then, is the minimum Cheeger ratio over all sets in the graph, $\varPhi(G) = \min_{S \subset V} \varPhi(S)$. The Cheeger ratio provides a quantitative measure concerning graph clusters and is related to the expansion and spectral gap of a graph [5].

Local Cluster and Sparse Cut. The sparse cut problem is to approximate the Cheeger ratio $\varPhi(G)$ of the graph. This is typically done by finding a set of vertices whose Cheeger ratio is close to $\varPhi(G)$– that is, a set which approximates the sparsest cut in the graph. For the local clustering problem, however, we are concerned with finding a set with small Cheeger ratio within a specified subset of vertices. Alternatively, one can view this as a sparse cut problem on an induced subgraph. This Cheeger ratio is sometimes called a *local Cheeger ratio* with respect to the specified subset.

A local clustering algorithm promises the following: Given a set S of Cheeger ratio ϕ, many vertices in S may serve as seeds for a sweep which finds a set of Cheeger ratio close to ϕ.

2.3   Our Results

In this work we give a distributed algorithm which computes a local cluster of Cheeger ratio $O(\sqrt{\phi})$ with high probability, while the optimal local cluster has Cheeger ratio ϕ. Our algorithm finishes in $O\left(\frac{\log(\epsilon^{-1})\log n}{\log\log(\epsilon^{-1})} + \frac{1}{\epsilon}\log n\right)$ rounds in the CONGEST model (Theorem 5) where ϵ is an error bound. Further, if ϕ is known, we show how to compute a local cluster in $O\left(\frac{\log(\epsilon^{-1})}{\log\log(\epsilon^{-1})} + \frac{1}{\epsilon}\right)$ rounds (Theorem 4). Our algorithm is an improvement of previous local clustering algorithms by eliminating a log factor in the performance guarantee. Further, its running time improves upon algorithms using standard and PageRank random walks. In particular, given the Cheeger ratio of an optimal local cluster, our algorithm runs in time only dependent upon the approximation error, ϵ, and is entirely independent of the input graph. The algorithms and accompanying analysis are given in Sect. 4.

Similar to existing local clustering algorithms, our algorithm uses a variation of random walks to compute a local cluster. However, rather than a standard random walk [25] or a PageRank random walk with reset probabilities [1], we use the *heat kernel random walk* (see Sect. 3).

We remark that in the analysis of random walks, the usual notion of approximation is total variation distance or some other vector norm based distance. However, in the approximation of PageRank or heat kernel pagerank for large graphs, the definition of approximation is quite different. Namely, we say some vector $\hat{\rho}_{t,s}$ is an ϵ-approximate PHKPR vector for $\rho_{t,s}$ with a seed vertex s and diffusion parameter $t \in R$ if:

1. $(1-\epsilon)\rho_{t,s}(v) - \epsilon \le \hat{\rho}_{t,s}(v) \le (1+\epsilon)\rho_{t,s}(v)$, and
2. for each node v with $\hat{\rho}_{t,s}(v) = 0$, it must be that $\rho_{t,s}(v) \le \epsilon$.

With the above definition of approximation, we here define the heat kernel pagerank approximation problem (or the *PHKPR problem* in short): given a vertex s of a graph and a diffusion parameter $t \in \mathbb{R}$, compute values $\hat{\rho}_{t,s}(v)$ for vertices v. We give a distributed algorithm which solves the PHKPR problem and finishes after only $O\left(\frac{\log(\epsilon^{-1})}{\log\log(\epsilon^{-1})}\right)$ rounds of communication (Theorem 2).

We extend our results to distributed k-machine model and show the existence of an algorithm which computes a local cluster over k machines in $\tilde{O}\left(\frac{\log(\epsilon^{-1})}{\epsilon^3 k^2 \log\log(\epsilon^{-1})} + \frac{1}{\epsilon k^2} + \left(\frac{\log(\epsilon^{-1})}{k\log\log(\epsilon^{-1})} + \frac{1}{k\epsilon}\right)\max\left\{\frac{1}{\epsilon^3}, \Delta\right\}\right)$ rounds, where Δ is the maximum degree in the graph, with high probability (Theorem 8). We note that when hiding polylogarithmic factors, this time does not depend on the size n of the graph. We compare this to an algorithm for computing a local cluster with PageRank which will require $\tilde{O}\left(\frac{\frac{1}{\alpha}+n}{k^2} + \left(\frac{1}{\alpha k} + \frac{n}{k}\right)\max\{\frac{1}{\epsilon}, \Delta\}\right)$ rounds with high probability, which is linear in n. These results are given in Sect. 5.

We briefly note here that local clustering algorithms can easily be extended to sparse cut algorithms. Namely, one can sample a number of random nodes in the network and perform the local clustering algorithm from each. One node in the network can store the Cheeger ratios output by each run of the algorithm and simply return the minimal Cheeger ratio as the value of the sparsest cut in the network. In [1, 25], $O(\frac{n}{\sigma} \log n)$ nodes are enough to compute a sparsest cut with high probability, where σ is the size of the cut set.

3   Fast Distributed Heat Kernel Pagerank Computation

The idea of the algorithm is to launch a number of random walks from the seed node in parallel, and compute the fraction of random walks which end at a node u as an estimate of the PHKPR values $\rho_{t,s}(u)$. Recall the definition of personalized heat kernel pagerank from (1), $\rho_{t,s} = \sum_{k=0}^{\infty} e^{-t} \frac{t^k}{k!} \chi_s P^k$. Then the values of this vector are exactly the stationary distribution of a *heat kernel random walk*: with probability $p_k = e^{-t} \frac{t^k}{k!}$, take k random walk steps according to the standard random walk transition probabilities P (see Sect. 2.2).

To be specific, the seed node s initializes r tokens, each of which holds a random variable k corresponding to the length of its random walk. Then, in rounds, the tokens are passed to random neighbors with a count incrementor until the count reaches k. At the end of the parallel random walks, each node holding tokens outputs the number of tokens it holds divided by r as an estimate for its PHKPR value. Algorithm 1 describes the full procedure.

Algorithm 1. DistributedEstimatePHKPR

---

input: a network modeled by a graph G, a seed node s, a diffusion parameter t, an error bound ϵ

output: estimates $\hat{\rho}_{t,s}(v)$ of PHKPR values for nodes v in the network

1: seed node s generates $r = \frac{16}{\epsilon^3} \log n$ tokens t_i
2: $K \leftarrow c \cdot \frac{\log(\epsilon^{-1})}{\log\log(\epsilon^{-1})}$ for any choice of $c \geq 1$
3: each token t_i does the following: pick a value k with probability $p_k = e^{-t} \frac{t^k}{k!}$, then hold the counter value $k_i \leftarrow \min\{k, K\}$
4: **for** iterations $j = 1 \ldots K$ **do**
5:     every node v performs the following in parallel:
6:     **for** every token t_i node v currently holds **do**
7:         **if** $k_i == j$ **then**
8:             hold on to this token for the duration of the iterations
9:         **else**
10:             send t_i to a random neighbor
11:         **end if**
12:     **end for**
13: **end for**
14: let C_v be the number of tokens node v currently holds
15: each node with $C_v > 0$ returns C_v/r as an estimate for its PHKPR value $\rho_{t,s}(v)$

The algorithm is based on that given in [6] in a static setting. Theorem 1 of [6] states that an ϵ-approximate PHKPR vector can be computed with the above procedure by setting $r = \frac{16}{\epsilon^3} \log n$. Further, the approximation guarantee holds when limiting the maximum length of random walks to $K = O\left(\frac{\log(\epsilon^{-1})}{\log\log(\epsilon^{-1})}\right)$, so that each token is passed for $\max\{k, K\}$ rounds, where k is drawn with probability p_k as described above. In the static setting, this limit keeps the running time down.

In contrast, the distributed algorithm DistributedEstimatePHKPR takes advantage of decentralized control to take multiple random walk steps via multiple edges at a time. That is, through parallel execution, the running time depends only on the length of random walks, whereas when running the random walks in serial, as in [6], the running time must also include the number of random walks performed. Thus, keeping K small is critical in keeping the number of rounds low, and is the key to the efficiency of our local clustering algorithms.

The correctness of the algorithm follows directly from Theorem 1 in [6], and is stated here without proof. The authors additionally give empirical evidence of the correctness of the algorithm with parameters $r = \frac{16}{\epsilon^3} \log n$ and $K = \frac{2\log(\epsilon^{-1})}{\log\log(\epsilon^{-1})}$ in an extended version of the paper [7]. They specifically demonstrate that the ranking of nodes obtained with an ϵ-approximate PHKPR vector computed this way is very close to the ranking obtained with an exact vector.

Theorem 1. *For any network G, any seed node $s \in V$, and any error bound $0 < \epsilon < 1$, the distributed algorithm DistributedEstimatePHKPR outputs an ϵ-approximate PHKPR vector with probability at least $1 - \epsilon$.*

The correctness of the algorithm holds for any choice of t, and in fact we use a particular value of t in our local clustering algorithm (see Sect. 4). Regardless, it is clear that the running time is independent of any choice of t. In fact, we demonstrate in the proof of Theorem 2 (given in the full version of this paper [8]) that it is independent of n as well.

Theorem 2. *For any network G, any seed node $s \in V$, and any error bound $0 < \epsilon < 1$, the distributed algorithm DistributedEstimatePHKPR finishes in $O\left(\frac{\log(\epsilon^{-1})}{\log\log(\epsilon^{-1})}\right)$ rounds.*

4   Distributed Local Cluster Detection

In this section we present a fast, distributed algorithm for the local clustering problem. The backbone of the algorithm involves investigating sets of nodes which accumulate in decreasing order of their $\hat{\rho}_{t,s}(v)/d_v$ values. The process is efficient and requires at most one linear scan of the nodes in the network (we actually show that the process can be much faster).

We describe the algorithm presently. Let p be any function over the nodes of the graph, and let π be the ordering of the nodes in decreasing order of $p(v)/d_v$. Then the majority of the work of the algorithm is investigating sufficiently many of the $n - 1$ cuts $(S_j, \bar{S}_j)$ given by the first j nodes in the ordering and the last $n - j$ nodes in the ordering, respectively, for $j = 1, \ldots, n$. However, by "sufficiently many" we indicate that we may stop investigating the cut sets when either the volume or the size of the set S_j is large. Assume this point is after $j = \mathbf{j}$. Then we choose the cut set that yields the minimum Cheeger ratio among the $\mathbf{j}$ possible cut sets. We call this process a *sweep*. As such, our local clustering algorithm is a *sweep* of a PHKPR distribution vector.

Algorithm 2. DistributedLocalCluster

---

input: a network modeled by a graph G, a seed node s, a target cluster size σ, a target cluster volume ς, an optimal Cheeger ratio ϕ, an error bound ϵ
output: a set of nodes S with $\Phi(S) \in O(\sqrt{\phi})$

1: $t \leftarrow \phi^{-1} \log(\frac{2\sqrt{\varsigma}}{1-\epsilon} + 2\epsilon\sigma)$
2: compute PHKPR values $\hat{\rho}_{t,s}(v)$ with DistributedEstimatePHKPR(G, s, t, ϵ)
3: every node v with a non-zero PHKPR value estimate sends $\hat{\rho}_{t,s}(v)/d_v$ to every other node with a non-zero PHKPR value estimate   ▷ *Phase 1*
4: let π be the ordering of nodes in decreasing order of $\hat{\rho}_{t,s}(v)/d_v$ ▷ *Phase 1*
5: compute Cheeger ratios of each of the cut sets with a call of the **Distributed sweep algorithm** and output the cut set of minimum Cheeger ratio ▷ *Phase 2*

In the static setting, this process will take $O(n \log n)$ time in general. The authors in [9] give a distributed sweep algorithm that finishes in $O(n)$ rounds. We improve the analysis of [9] using a PHKPR vector. The running time of our sweep algorithm is given in Lemma 1, and the proof is provided in [8].

The sweep involves two phases. In Phase 1, the goal is for each node to know its place in the ordering π. Each node can compute their own $\hat{\rho}_{t,s}(v)/d_v$ value locally, and we use $O(\frac{1}{\epsilon})$ rounds to ensure each node knows the π values of all other nodes (see the proof of Lemma 1 [8]). In Phase 2, we use the decentralized sweep of [9] described presently:

Distributed Sweep Algorithm. Let N denote the number of nodes with a non-zero estimated PHKPR value after running the algorithm DistributedEstimatePHKPR. Assume each node knows its position in ordering π after Phase 1. We will refer to nodes by their place in the ordering. Define S_j to be the cut set of the first j nodes in the ordering. Then computing the Cheeger ratio of each cut set S_j involves a computation of the volume of the set as well as $|E(S_j, \bar{S}_j)|$. Define the following:

- L_j^π is the number of neighbors of node j in S_{j-1}, and
- R_j^π is the number of neighbors of node j in $\bar{S}_j$.

Then the Cheeger ratio of each cut set can be computed locally by:

$$\circ\ |E(S_j, \bar{S}_j)| = |E(S_{j-1}, \bar{S}_{j-1})| - L_j^\pi + R_j^\pi, \text{ with } |E(S_1, \bar{S}_1)| = d_1 \quad (2)$$

$$\circ\ \mathrm{vol}(S_j) = \mathrm{vol}(S_{j-1}) + L_j^\pi + R_j^\pi, \text{ with } vol(S_1) = d_1. \quad (3)$$

We now show that a sweep can be performed in $O(N)$ rounds. Each node knows the IDS of its neighbors and after Phase 1 each node knows the place of every other node in the ordering π. Therefore, each node can compute locally if a neighbor is in S_{j-1} or $\bar{S}_j$, and so L_j^π and R_j^π can be computed locally for each node j. Each node can then prepare an $O(\log n)$-bit message of the form $(\text{ID}, L_j^\pi, R_j^\pi)$. Each of the N messages of this form can then be sent to the first node in the ordering using the upcasting algorithm (described in the proof of Lemma 1) using the π ordering as node rank. We note that the N nodes in the ordering are necessarily in a connected component of the network, and so the upcasting procedure can be performed in $O(N)$ rounds. Finally, once the first node in the ordering is in possession of the ordering π, and the values of (L_j^π, R_j^π) for every node in the ordering, it may iteratively compute $\Phi(S_j)$ locally using the rules (2) and (3). Thus, this node can output the minimum Cheeger ratio ϕ^* as well as the j^* such that $\Phi(S_{j^*}) = \phi^*$ after $O(N)$ rounds.

Lemma 1. *Performing Phases 1 and 2 of a distributed sweep takes $O(\frac{1}{\epsilon})$ rounds.*

The algorithm DistributedLocalCluster (Algorithm 2) is a complete description of our distributed local clustering algorithm. The correctness of the algorithm follows directly from [6] and we omit the proof here.

Theorem 3. *For any network G, suppose there is a set of Cheeger ratio ϕ. Then at least half of the vertices in S can serve as the seed s so that for any error bound $0 < \epsilon < 1$, the algorithm DistributedLocalCluster will find a set of Cheeger ratio $O(\sqrt{\phi})$ with probability at least $1 - \epsilon$.*

Theorem 4. *For any network G, any seed node $s \in V$, and any error bound $0 < \epsilon < 1$, the algorithm DistributedLocalCluster finishes in $O\left(\frac{\log(\epsilon^{-1})}{\log\log(\epsilon^{-1})} + \frac{1}{\epsilon}\right)$ rounds.*

Proof. The only distributed computations are those for computing approximate PHKPR values (line 2) and Phase 1 (lines 3 and 4) and Phase 2 (line 5) of the distributed sweep. Computing PHKPR values takes $O\left(\frac{\log(\epsilon^{-1})}{\log\log(\epsilon^{-1})}\right)$ rounds by Theorem 2, and Phases 1 and 2 together take $O(\frac{1}{\epsilon})$ rounds by Lemma 1. Thus the running time follows.

One possible concern with the algorithm DistributedLocalCluster is that one cannot guarantee knowing the value of ϕ ahead of time for any particular node s. Therefore a true local clustering algorithm should be able to proceed without this information. This can be achieved by "testing" a few values of ϕ (and fixing some reasonable values for σ and ς). Namely, begin with $\phi = 1/2$ and run the algorithm above. If the output cut set S satisfies $\Phi(S) \in O(\sqrt{\phi})$, we are done. If not, halve the value of ϕ and continue. There are $O(\log n)$ such guesses, and we have arrived at the following.

Theorem 5. *For any network G, any node s, and any error bound $0 < \epsilon < 1$, there is a distributed algorithm that computes a set S with Cheeger ratio within a quadratic of the optimal which finishes in $O\left(\frac{\log(\epsilon^{-1})\log n}{\log\log(\epsilon^{-1})} + \frac{1}{\epsilon}\log n\right)$ rounds.*

In particular, when ignoring polylogarithmic factors, the running time is $\tilde{O}\left(\frac{\log(\epsilon^{-1})}{\log\log(\epsilon^{-1})} + \frac{1}{\epsilon}\right)$.

5   Computing Local Clusters in the k-machine Model

In this section we consider a graph on n vertices which is distributed across k nodes in a computer network. This is the k-machine model introduced in Sect. 2.1.

In the k-machine model, we consider a network of $k > 1$ distinct machines that are pairwise interconnected by bidirectional point-to-point communication links. Each machine executes an instance of a distributed algorithm. The computation advances in rounds where, in each round, machines can exchange messages through their communication links. We again assume that each link has a bandwidth of $O(\log n)$ meaning that $O(\log n)$ bits may be transmitted through a link in any round. We also assume no shared memory and no other means of communication between nodes. When we say an algorithm solves a problem in x rounds, we mean that x is the maximum number of rounds until termination of the algorithm, over all n-node, m-edge graphs G.

In this model we are solving massive graph problems in which the vertices of the graph are distributed among the k machines. We assume $n \geq k$ (typically $n \gg k$). Initially the entire graph is not known by a single machine but rather partitioned among the k machines in a "balanced" fashion so that the nodes and/or edges are partitioned approximately evenly among the machines. There are several ways of partitioning vertices, and we will consider a random partition, where vertices and incident edges are randomly assigned to machines. Formally, each vertex v of G is assigned independently and randomly to one of the k machines, which we call the home machine of v. The home machine of v thereafter knows the ID of v as well as the IDs and home machines of neighbors of v.

In the remainder of this section we prove the existence of efficient algorithms for solving the PHKPR and local cluster problems in the k-machine model. Our main tool is the Conversion Theorem of [14].

Define M as the *message complexity*, the worst case number of messages sent in total during a run of the algorithm. Also define C as the *communication degree complexity*, or the maximum number of messages sent or received by any node in any round of the algorithm. Then we use as a key tool the Conversion Theorem as restated below.

Theorem 6 (Conversion Theorem [14]). *Suppose there is an algorithm A_C that solves problem P in the CONGEST model for any n-node graph G with probability at least $1-\epsilon$ in time $T_C(n)$. Further, let A_C use message complexity M and communication degree complexity C. Then there exists an algorithm A_k that solves P for any n-node graph G with probability at least $1 - \epsilon$ in the k-machine model in $\tilde{O}\left(\frac{M}{k^2} + \frac{T_C(n)C}{k}\right)$ rounds with high probability.*

In the forthcoming theorems, by "high probability" we mean with probability at least $1 - 1/n$.

We note that the proof of the Conversion Theorem is constructive, describing precisely how an algorithm A_k in the k-machine model simulates the algorithm A_C in the CONGEST model. We omit the simulation here but encourage the reader to refer to the proof for implementation details.

By Theorem 2, we know that PHKPR values can be estimated with ϵ-accuracy in $O\left(\frac{\log(\epsilon^{-1})}{\log\log(\epsilon^{-1})}\right)$ rounds. A total of $O\left(\frac{1}{\epsilon^3}\log n\right)$ messages are generated and propogated for at most $O\left(\frac{\log(\epsilon^{-1})}{\log\log(\epsilon^{-1})}\right)$ random walk steps, for a total of $O\left(\frac{\log(\epsilon^{-1})\log n}{\epsilon^3\log\log(\epsilon^{-1})}\right)$ messages sent during a run of the algorithm. In the first random walk step, each of the $O\left(\frac{1}{\epsilon^3}\log n\right)$ messages may be passed to a neighbor of the seed node, so the message complexity is $O\left(\frac{1}{\epsilon^3}\log n\right)$. Therefore we arrive at the following.

Theorem 7. *There exists an algorithm that solves the PHKPR problem for any n-node graph in the k-machine model with probability at least $1-\epsilon$ and runs in $\tilde{O}\left(\frac{\log(\epsilon^{-1})}{\epsilon^3 k\log\log(\epsilon^{-1})}\left(\frac{1}{k}+1\right)\right)$ rounds with high probability.*

By Theorem 5, a local cluster about any seed node can be computed in $O\left(\frac{\log(\epsilon^{-1})\log n}{\log\log(\epsilon^{-1})}+\frac{1}{\epsilon}\log n\right)$ rounds. The message complexity for the PHKPR phase is $O\left(\left(\frac{\log(\epsilon^{-1})\log n}{\epsilon^3\log\log(\epsilon^{-1})}\right)\log n\right)$ and for the sweep phase is $O\left(\frac{1}{\epsilon}\log n\right)$, for a total message complexity of $O\left(\frac{\log(\epsilon^{-1})\log^2 n}{\epsilon^3\log\log(\epsilon^{-1})}+\frac{1}{\epsilon}\log n\right)$. The communication degree complexity is $O\left(\frac{1}{\epsilon^3}\log n\right)$ for the PHKPR phase (as above), and $O(\Delta)$, where Δ is the maximum degree in the graph, for the sweep phase. Thus the communication degree complexity for the algorithm is the maximum of these two. We therefore have the following result for the k-machine model.

Theorem 8. *There exists an algorithm that computes a local cluster for any n-node graph in the k-machine model with probability at least $1-\epsilon$ and runs in $\tilde{O}\left(\frac{\log(\epsilon^{-1})}{\epsilon^3 k^2\log\log(\epsilon^{-1})}+\frac{1}{\epsilon k^2}+\left(\frac{\log(\epsilon^{-1})}{k\log\log(\epsilon^{-1})}+\frac{1}{k\epsilon}\right)\max\left\{\frac{1}{\epsilon^3},\Delta\right\}\right)$ rounds, where Δ is the maximum degree in the graph, with high probability.*

Acknowledgements. The authors would like to warmly thank Yiannis Koutis for discussion and for suggesting the problem of finding efficient distributed algorithms, as well as the anonymous reviewers for their suggestions for improving the paper.

References

1. Andersen, R., Chung, F., Lang, K.: Local graph partitioning using pagerank vectors. In: FOCS, pp. 475–486. IEEE (2006)
2. Andersen, R., Peres, Y.: Finding sparse cuts locally using evolving sets. In: STOC, pp. 235–244. ACM (2009)
3. Arora, S., Rao, S., Vazirani, U.: Expander flows, geometric embeddings and graph partitioning. JACM **56**(2), 1–37 (2009). Article no. 5

4. Brin, S., Page, L.: The anatomy of a large-scale hypertextual web search engine. Comput. Netw. ISDN Syst. **30**(1), 107–117 (1998)
5. Chung, F.: Spectral Graph Theory. American Mathematical Society, Providence (1997)
6. Chung, F., Simpson, O.: Computing heat kernel pagerank and a local clustering algorithm. In: Jan, K., Miller, M., Froncek, D. (eds.) IWOCA 2014. LNCS, vol. 8986, pp. 110–121. Springer, Heidelberg (2015)
7. Chung, F., Simpson, O.: Computing heat kernel pagerank and a local clustering algorithm. arXiv preprint arXiv:1503.03155 (2015)
8. Chung, F., Simpson, O.: Distributed algorithms for finding local clusters using heat kernel pagerank. arXiv preprint arXiv:1507.08967 (2015)
9. Das Sarma, A., Molla, A.R., Pandurangan, G.: Distributed computation of sparse cuts via random walks. In: ICDCN, pp. 6:1–6:10 (2015)
10. Das Sarma, A., Molla, A.R., Pandurangan, G., Upfal, E.: Fast distributed pagerank computation. In: Frey, D., Raynal, M., Sarkar, S., Shyamasundar, R.K., Sinha, P. (eds.) ICDCN 2013. LNCS, vol. 7730, pp. 11–26. Springer, Heidelberg (2013)
11. Das Sarma, A., Nanongkai, D., Pandurangan, G., Tetali, P.: Distributed random walks. JACM **60**(1), 201–210 (2013). Article no. 2
12. Dean, J., Ghemawat, S.: Mapreduce: simplified data processing on large clusters. In: OSDI (2004)
13. Gharan, S.O., Trevisan, L.: Approximating the expansion profile and almost optimal local graph clustering. In: FOCS, pp. 187–196. IEEE (2012)
14. Klauck, H., Nanongkai, D., Pandurangan, G., Robinson, P.: Distributed computation of large-scale graph problems. In: SODA, pp. 391–410. SIAM (2015)
15. Kloster, K., Gleich, D.F.: Heat kernel based community detection. In: ACM SIGKDD, pp. 1386–1395. ACM (2014)
16. Leskovec, J., Lang, K.J., Dasgupta, A., Mahoney, M.W.: Statistical properties of community structure in large social and information networks. In: WWW, pp. 695–704. ACM (2008)
17. Liao, C.S., Lu, K., Baym, M., Singh, R., Berger, B.: Isorankn: spectral methods for global alignment of multiple protein networks. Bioinformatics **25**(12), i253–i258 (2009)
18. Lovász, L., Simonovits, M.: The mixing rate of markov chains, an isoperimetric inequality, and computing the volume. In: FOCS, pp. 346–354. IEEE (1990)
19. Lovász, L., Simonovits, M.: Random walks in a convex body and an improved volume algorithm. Random Struct. Algorithms **4**(4), 359–412 (1993)
20. Low, Y., Gonzalez, J., Kyrola, A., Bickson, D., Guestrin, C., Hellerstein, J.M.: Graphlab: a new framework for parallel machine learning. In: UAI, pp. 340–349 (2010)
21. Malewicz, G., Austern, M.H., Bik, A.J., Dehnert, J.C., Horn, I., Leiser, N., Czajkowski, G.: Pregel: a system for large-scale graph processing. In: SIGMOD International Conference on Management of data, pp. 135–146. ACM (2010)
22. Orecchia, L., Sachdeva, S., Vishnoi, N.K.: Approximating the exponential, the lanczos method and an $\tilde{O}(m)$-time spectral algorithm for balanced separator. In: STOC, pp. 1141–1160. ACM (2012)
23. Pandurangan, G., Khan, M.: Theory of communication networks. In: Atallah, M.J., Blanton, M. (eds.) Algorithms and Theory of Computation Handbook. Chapman & Hall/CRC, Boca Raton (2010)
24. Peleg, D.: Distributed computing. In: SIAM Monographs on Discrete Mathematics and Applications 5 (2000)
25. Spielman, D.A., Teng, S.H.: Nearly-linear time algorithms for graph partitioning, graph sparsification, and solving linear systems. In: STOC, pp. 81–90. ACM (2004)

Strong Localization in Personalized PageRank Vectors

Huda Nassar[1]([⊠]), Kyle Kloster[2], and David F. Gleich[1]

[1] Computer Science Department, Purdue University, West Lafayette, USA
{hnassar,dgleich}@purdue.edu
[2] Mathematics Department, Purdue University, West Lafayette, USA
kkloste@purdue.edu

Abstract. The personalized PageRank diffusion is a fundamental tool in network analysis tasks like community detection and link prediction. It models the spread of a quantity from a set of seed nodes, and it has been observed to stay localized near this seed set. We derive an upper-bound on the number of entries necessary to approximate a personalized PageRank vector in graphs with skewed degree sequences. This bound shows localization under mild assumptions on the maximum and minimum degrees. Experimental results on random graphs with these degree sequences show the bound is loose and support a conjectured bound.

Keywords: PageRank · Diffusion · Local algorithms

1 Introduction

Personalized PageRank vectors [23] are a ubiquitous tool in data analysis of networks in biology [12,21] and information-relational domains such as recommender systems and databases [15,17,22]. In contrast to the standard PageRank vector, personalized PageRank vectors model a random-walk process on a network that randomly returns to a fixed starting node instead of restarting from a random node in the network as in the traditional PageRank. This process is also called a random-walk with restart.

The stationary distributions of the resulting process are typically called personalized PageRank vectors. We prefer the terms "localized PageRank" or "seeded PageRank" as these choices are not as tied to PageRank's origins on the web. A seeded PageRank vector depends on three terms: the network modeled as a column-stochastic matrix $\mathbf{P}$ characterizing the random-walk process, a parameter α that determines the restart probability $(1 - \alpha)$, and a seed node s. The vector $\mathbf{e}_s$ is the vector of all zeros with a single 1 in the position corresponding to node s. The seeded PageRank vector $\mathbf{x}$ is then the solution of the linear system:

$$(\mathbf{I} - \alpha\mathbf{P})\mathbf{x} = (1 - \alpha)\mathbf{e}_s.$$

K. Kloster and D.F. Gleich—Supported by NSF CAREER award CCF-1149756 and DARPA SIMPLEX Code available online https://github.com/nassarhuda/pprlocal.

D.F. Gleich et al. (Eds.): WAW 2015, LNCS 9479, pp. 190–202, 2015.
DOI: 10.1007/978-3-319-26784-5_15

When the network is strongly connected, the solution $\mathbf{x}$ is non-zero for all nodes. This is because there is a non-zero probability of walking from the seed to any other node in a strongly connected network. Nevertheless, the solution $\mathbf{x}$ displays a behavior called *localization*. We can attain accurate localized PageRank solutions by truncating small elements of $\mathbf{x}$ to zero. Put another way, there is a sparse vector $\mathbf{x}_\varepsilon$ that approximates $\mathbf{x}$ to an accuracy of ε. This behavior is desirable for applications of seeded PageRank because they typically seek to "highlight" a small region related to the seed node s inside a large graph.

The essential question we study in this paper is: how sparse can we make $\mathbf{x}_\varepsilon$? To be precise, we consider a notion of strong localization, $\|\mathbf{x}_\varepsilon - \mathbf{x}\|_1 \leq \varepsilon$, and we focus on the behavior of $f(\varepsilon) := \min \mathrm{nonzeros}(\mathbf{x}_\varepsilon)$. Note that $\mathbf{x}_\varepsilon$ depends on α, the particular random-walk on the graph $\mathbf{P}$, and the seed node s from which the PageRank diffusion begins. We only consider stochastic matrices $\mathbf{P}$ that arise from random-walks on strongly-connected graphs. So a more precise statement of our goal is:

$$f_\alpha(\varepsilon) = \max_{\mathbf{P}} \max_{s} \min_{\mathbf{x}_\varepsilon} \mathrm{nonzeros}(\mathbf{x}_\varepsilon) \text{ where } \|\mathbf{x}_\varepsilon - \mathbf{x}(\alpha, \mathbf{P}, s)\|_1 \leq \varepsilon,$$

and where $\mathbf{x}(\alpha, \mathbf{P}, s)$ is the seeded PageRank vector $(1 - \alpha)(\mathbf{I} - \alpha\mathbf{P})^{-1}\mathbf{e}_s$. The goal is to establish bounds on $f(\varepsilon)$ that are sublinear in n, the number of nodes of the graph, because that implies localized solutions to PageRank.

Adversarial localized PageRank constructions exist where the solutions $\mathbf{x}$ are near the uniform distribution (see Sect. 2). Thus, it is not possible to meaningfully bound $f(\varepsilon)$ as anything other than n. It is also known that $f(\varepsilon)$ is sublinear in n for graphs that are essentially of bounded maximum degree [6] due to resolvent theory. The case of skewed degrees was open until our result.

We establish an upper-bound on $f_\alpha(\varepsilon)$ as a function of the rate of decay of the degree sequence, $1/\varepsilon$, α, the maximum degree d, and the minimum degree δ (Theorem 1). This bound enables us to establish sublinear localization for graphs with growing maximum degrees provided that the other node degrees decay sufficiently rapidly. When we study this bound in random realizations of appropriate networks, it turns out to be loose; hence, we develop a new conjectured bound (Sect. 4).

1.1   Related Work on Weak Localization

There is another notion of localization that appears in uses of PageRank for partitioning undirected graphs:

$$\|\mathbf{D}^{-1}(\mathbf{x}_\varepsilon - \mathbf{x})\|_\infty = \max_{i} |[x_\varepsilon]_i - x_i|/d_i \leq \varepsilon.$$

If this notion is used for a localized Cheeger inequality [1,10], then we need the additional property that $0 \leq \mathbf{x}_\varepsilon \leq \mathbf{x}$ element-wise. When restated as a localization result, the famous Andersen-Chung-Lang PageRank partitioning result [1] includes a proof that:

$$\max_{\mathbf{P}} \max_{s} \min_{\mathbf{x}_\varepsilon} \mathrm{nonzeros}(\mathbf{x}_\varepsilon) \leq \frac{1}{1-\alpha}\frac{1}{\varepsilon}, \text{ where } \|\mathbf{D}^{-1}(\mathbf{x}_\varepsilon - \mathbf{x}(\alpha, \mathbf{P}, s))\|_\infty \leq \varepsilon.$$

This establishes that *any* uniform random walk on a graph satisfies a weak-localization property. The paper also gives a fast algorithm to find these weakly local solutions. More recently, there have appeared a variety of additional weak-localization results on diffusions [13,19].

1.2   Related Work on Functions of Matrices and Diffusions

Localization in diffusions is broadly related to localization in functions of matrices [6]. The results in that literature tend to focus on the case of banded matrices (e.g. [5]), although there are also discussions of more general results in terms of graphs arising from sparse matrices [6]. These same types of decay bounds can apply to a variety of graph diffusion models that involve a stochastic matrix [3,16], and recent work shows that they may even extend beyond this regime [13]. In the context of the decay of functions of matrices, we advance the literature by proving a localization bound for a particular resolvent function of a matrix that applies to graphs with growing maximum degree.

2   A Negative Result for Strong Localization

Here we give an example of a graph that always has a non-local seeded PageRank vector. More concretely, we demonstrate the existence of a personalized PageRank vector that requires $\Theta(n)$ nonzeros to attain a 1-norm accuracy of ε, where n is the number of nodes in the graph.

The graph is just the undirected star graph on n nodes. Then the PageRank vector $\mathbf{x}$ seeded on the center node has value $1/(1 + \alpha)$ for the center node and $\alpha/((1+\alpha)(n-1))$ for all leaf nodes. Suppose an approximation $\hat{\mathbf{x}}$ of $\mathbf{x}$ has M of these leaf-node entries set to 0. Then the 1-norm error $\|\mathbf{x} - \hat{\mathbf{x}}\|_1$ would be at least $M\alpha/((1+\alpha)(n-1))$. Attaining a 1-norm accuracy of ε requires $M\alpha/((1+\alpha)(n-1)) < \varepsilon$, and so the minimum number of entries of the approximate PageRank vector required to be non-zero $(n - M)$ is then lower-bounded by $n(1 - c) + c$, where $c = \varepsilon(1 + \alpha)/\alpha$. Note that this requires $c \in (0, 1)$, which holds if $\varepsilon < \alpha/2$. Thus, the number of nonzeros required in the approximate PageRank vector must be linear in n.

3   Localization in Personalized PageRank

The example in Sect. 2 demonstrates that there exist seeded PageRank vectors that are *non-local*. Here we show that graphs with a particular type of skewed degree sequence and a growing, but sublinear, maximum degree have seeded PageRank vectors that are always localized, and we give an upper-bound on $f(\varepsilon)$ for this class of graph. This theorem originates in our recent work on seeded heat kernel vectors [14], and we now employ similar arguments to treat seeded PageRank vectors. Our present analysis yields tighter intermediate inequalities and results in an entirely novel bound for the localization of PageRank.

Theorem 1. *Let* $\mathbf{P}$ *be a uniform random walk transition matrix of a graph on* n *nodes with maximum degree* d *and minimum degree* δ. *Additionally, suppose that the* k*th largest degree,* $d(k)$, *satisfies* $d(k) \leq \max\{dk^{-p}, \delta\}$. *The Gauss-Southwell coordinate relaxation method applied to the seeded PageRank problem* $(\mathbf{I} - \alpha\mathbf{P})\mathbf{x} = (1 - \alpha)\mathbf{e}_s$ *produces an approximation* $\boldsymbol{x}_{\varepsilon}$ *satisfying* $\|\boldsymbol{x} - \boldsymbol{x}_{\varepsilon}\|_1 < \varepsilon$ *having at most* N *non-zeros in the solution, where* N *satisfies*

$$N = \min\left\{ n, \; \tfrac{1}{\delta}C_p\left(\tfrac{1}{\varepsilon}\right)^{\frac{\delta}{1-\alpha}} \right\}, \tag{1}$$

and where we define C_p *to be*

$$\begin{aligned} C_p &:= d(1 + \log d) && \textit{if } p = 1 \\ &:= d\left(1 + \tfrac{1}{1-p}\left(d^{\frac{1}{p}-1} - 1\right)\right) && \textit{otherwise.} \end{aligned}$$

Note that the upper bound $N = n$ is trivial as a vector cannot have more non-zeros than entries. Thus, d, δ, p, and n must satisfy certain conditions to ensure that inequality (1) is not trivial. In particular, for values of $p < 1$, it is necessary that $d = o(n^p)$ for inequality (1) to imply that $N = o(n)$. For $p > 1$, the bound guarantees sublinear growth of N as long as $d = o(n)$. Additionally, the minimum degree δ must be bounded by $O(\log \log n)$. Thus we arrive at:

Corollary 1. *Let* G *be a class of graphs with degree sequences obeying the conditions of Theorem 1 with constant* δ *and* $d = o(n^{\min(p,1)})$. *Then* $f(\varepsilon) = o(n)$, *and seeded PageRank vectors are localized.*

We also note that the theorem implies localized seeded PageRank vectors for any graph with a maximum degree $d = O(\log \log n)$.

3.1   Our Class of Skewed Degree Sequences

We wish to make a few remarks about the class of skewed degree sequences where our results apply. Perhaps the most well-known is the power-law degree distribution where the probability that a node has degree k is proportional to $k^{-\gamma}$. These power-laws can be related to our skewed sequences with $p = 1/(\gamma-1)$ and $d = O(n^p)$ [2]. This setting renders our bound trivial with n nonzeros. Nevertheless, there is evidence that some real-world networks exhibit our type of skewed degrees [11] where the bound is asymptotically non-trivial.

3.2   Deriving the Bound

Getting back to the proof, our goal is an ε-approximation, $\mathbf{x}_\varepsilon$, to the equation $(\mathbf{I} - \alpha\mathbf{P})\mathbf{x} = (1 - \alpha)\mathbf{e}_s$ for a seed s. Given an approximation, $\hat{\mathbf{x}}$, we can express the error in terms of the residual vector $\mathbf{r} = (1 - \alpha)\mathbf{e}_s - (\mathbf{I} - \alpha\mathbf{P})\hat{\mathbf{x}}$ as follows:

$$\mathbf{x} - \hat{\mathbf{x}} = (\mathbf{I} - \alpha\mathbf{P})^{-1}\mathbf{r}. \tag{2}$$

Using this relationship, we can bound our approximation's 1-norm accuracy, $\|\mathbf{x} - \hat{\mathbf{x}}\|_1$, with the quantity $\frac{1}{1-\alpha}\|\mathbf{r}\|_1$. This is because the column-stochasticity of $\mathbf{P}$ implies that $\|(\mathbf{I} - \alpha\mathbf{P})^{-1}\|_1 = \frac{1}{1-\alpha}$. Guaranteeing a 1-norm error $\|\mathbf{x} - \hat{\mathbf{x}}\|_1 < \varepsilon$ is then a matter of ensuring that $\|\mathbf{r}\|_1 < (1-\alpha)\varepsilon$ holds. To bound the residual norm, we look more closely at a particular method for producing the approximation.

The Gauss-Southwell iteration. The Gauss-Southwell algorithm is a coordinate relaxation method for solving a linear system akin to the Gauss-Seidel linear solver. When solving a linear system, the Gauss-Southwell method proceeds by updating the entry of the approximate solution that corresponds to the largest magnitude entry of the residual, $\mathbf{r}$. We describe the Gauss-Southwell update as it is used to solve the seeded PageRank linear system.

The algorithm begins by setting the initial solution $\mathbf{x}^{(0)} = 0$ and $\mathbf{r}^{(0)} = (1-\alpha)\mathbf{e}_s$. In step k, let $j = j(k)$ be the entry of $\mathbf{r}^{(k)}$ with the largest magnitude, and let $m = |\mathbf{r}_j^{(k)}|$. We update the solution $\mathbf{x}^{(k)}$ and residual as follows:

$$\mathbf{x}^{(k+1)} = \mathbf{x}^{(k)} + m\mathbf{e}_j \tag{3}$$

$$\mathbf{r}^{(k+1)} = \mathbf{e}_s - (\mathbf{I} - \alpha\mathbf{P})\mathbf{x}^{(k+1)}, \tag{4}$$

and the residual update can be expanded to $\mathbf{r}^{(k+1)} = \mathbf{r}^{(k)} - m\mathbf{e}_j + m\alpha\mathbf{P}\mathbf{e}_j$. Since each update to the solution $\mathbf{x}^{(k)}$ alters exactly one entry of the vector, the index k is an upper-bound on the number of non-zeros in the solution.

This application of Gauss-Southwell to seeded PageRank-style problems has appeared numerous times in recent literature [7,8,18,20]. In at least one instance ([8], Sect. 5.2) the authors showed that the residual and solution vector stay nonnegative throughout this process, assuming the seed vector is nonnegative (which, in our context, it is). So the 1-norm of the residual can be expressed as $\|\mathbf{r}^{(k+1)}\|_1 = \mathbf{e}^T\mathbf{r}^{(k+1)}$, where $\mathbf{e}$ is the vector of all ones. Expanding the residual in terms of the iterative update presented above, we can write the residual norm as $\mathbf{e}^T\left(\mathbf{r}^{(k)} - m\mathbf{e}_j + m\alpha\mathbf{P}\mathbf{e}_j\right)$. Then, denoting $\|\mathbf{r}^{(k)}\|_1$ by r_k, yields the recurrence $r_{k+1} = r_k - m(1-\alpha)$.

Next observe that since m is the largest magnitude entry in $\mathbf{r}$, it is larger than the average value of $\mathbf{r}$. Let $Z(k)$ denote the number of nonzero entries in $\mathbf{r}^{(k)}$; then the average value can be expressed as $r_k/Z(k)$. Hence, we have $m \geq r_k/Z(k)$, and so we can bound $r_k - m(1-\alpha)$ above by $r_k - r_k(1-\alpha)/Z(k)$. Thus, $r_{k+1} \leq r_k\left(1 - (1-\alpha)/Z(k)\right)$, and we can recur to find:

$$r_{k+1} \leq r_0 \prod_{t=0}^{k}\left(1 - \tfrac{1-\alpha}{Z(t)}\right), \tag{5}$$

where $r_0 = (1-\alpha)$ because $\mathbf{r}_0 = (1-\alpha)\mathbf{e}_s$. Then, using the fact that $\log(1-x) \leq -x$ for $x < 1$, we note:

$$r_{k+1} \leq (1-\alpha) \prod_{t=0}^{k}\left(1 - \tfrac{1-\alpha}{Z(t)}\right) \leq (1-\alpha)\exp\left(-(1-\alpha)\sum_{t=0}^{k}\tfrac{1}{Z(t)}\right). \tag{6}$$

To progress from here we need some control over the quantity $Z(t)$ and this is where our skewed degree sequence enters the proof.

3.3   Using the degree sequence

We show that for a graph with this kind of skewed degree sequence, the number of entries in the residual obeys:

$$Z(t) \le C_p + \delta t, \tag{7}$$

where the term C_p is defined in the statement of Theorem 1. A similar analysis was presented in [14], but the current presentation improves the bound on C_p. This bound is proved below, but first we use this bound on $Z(t)$ to control the bound on r_k. Lemma 5.6 from [14] implies that

$$\sum_{t=0}^{k} \frac{1}{Z(t)} \ge \frac{1}{\delta} \log \left((\delta(k+1) + C_p)/C_p \right)$$

and so, plugging into (6), we can bound

$$r_{k+1} \le (1-\alpha) \exp \left(-\frac{(1-\alpha)}{\delta} \log \left(\frac{(\delta(k+1)+C_p)}{C_p} \right) \right),$$

which simplifies to $r_{k+1} \le (1-\alpha) \left((\delta(k+1) + C_p)/C_p \right)^{(\alpha-1)/\delta}$. Finally, to guarantee $r_k < \varepsilon(1-\alpha)$, it suffices to choose k so that $\left((\delta k + C_p)/C_p \right)^{(\alpha-1)/\delta} \le \varepsilon$. This holds if and only if $(\delta k + C_p) \ge C_p (1/\varepsilon)^{\delta/(\alpha-1)}$ holds, which is guaranteed by $k \ge \frac{1}{\delta} C_p (1/\varepsilon)^{\delta/(1-\alpha)}$. Thus, $k = \frac{1}{\delta} C_p (1/\varepsilon)^{\delta/(1-\alpha)}$ steps will produce an ε-approximation. Each step introduces at most one non-zero, which implies that if $k < n$, then there is an approximation $\mathbf{x}_\varepsilon$ with $N = k < n$ non-zeros. If $k \ge n$, then this analysis produces the trivial bound $N = n$.

Proving the degree sequence bound. Here we prove the inequality in (7) used in the proof above. For additional details, see the proof of Lemma 5.5 in [14], which is similar but results in a slightly worse bound. First, observe that the number of nonzeros in the residual after t steps is bounded above by the sum of the largest t degrees, $Z(t) \le \sum_{k=1}^{t} d(k)$. When we substitute the decay bound $d(k) \le dk^{-p}$ into this expression, $d(k)$ is only a positive integer when $k \le (d/\delta)^{1/p}$. Hence, we split the summation $Z(t) \le \sum_{k=1}^{t} d(k)$ into two pieces,

$$Z(t) \quad \le \quad \sum_{k=1}^{t} d(k) \quad \le \quad \left(\sum_{k=1}^{\lfloor (d/\delta)^{1/p} \rfloor} dk^{-p} \right) + \sum_{k=\lfloor (d/\delta)^{1/p} \rfloor + 1}^{t} \delta.$$

We want to prove that this implies $Z(t) \le C_p + \delta t$. The second summand is always less than δt. The first summand can be bounded above by $d \left(1 + \int_1^{(d/\delta)^{1/p}} x^{-p} dx \right)$ using a right-hand integral rule. This integral is straightforward to bound above with the quantity C_p defined in Theorem 1. This completes the proof.

4   Experiments

We present experimental results on the localization of seeded PageRank vectors on random graphs that have our skewed degree sequence and compare the actual sparsity with the predictions of our theoretical bound. This involves generating random graphs with the given skewed degree sequence (Sect. 4.1) and then comparing the experimental localization with our theoretical bound (Sect. 4.3). The bound is not particularly accurate, and so we conjecture a new bound that better predicts the behavior witnessed (Sect. 4.4).

4.1   Generating the Graphs

For experimental comparison, we wanted a test suite of graphs with varying but specific sizes and degree sequences. To produce these graphs, we use the Bayati-Kim-Saberi procedure [4] for generating undirected graphs with a prescribed degree sequences. The degree sequences used follow our description in Theorem 1 precisely. We choose the maximum degree d to be $n^{1/3}$ or $n^{1/2}$ and the minimum degree to be $\delta = 2$. We use several values for the decay exponent p, stated below.

After generating the degree sequence, we use the Erdős-Gallai conditions and the Havel-Hakimi algorithm to check if it is graphical. If the previously generated sequence fails, we perturb the sequence slightly and recheck the conditions. It often fails because the sequence has an odd sum; to resolve this state it suffices to increase the degree of one of the nodes with minimum degree by 1. Lastly, we verify that the graph contains a large connected component. We proceed once a graph has been generated that meets the above conditions and has a largest connected component that includes at least $n(1 - 10^{-2})$ nodes.

4.2   Measuring the Non-zeros

Given a graph, we first use the power method to compute a PageRank vector, seeded on the node with largest degree, to high-accuracy (1-norm error bounded by 10^{-12}). This requires $\lfloor (\log(\varepsilon/2))/(\log(\alpha)) \rfloor$ iterations based on the geometric convergence rate of α. We then study vectors $\mathbf{x}_\varepsilon$ satisfying $\|\mathbf{x}_\varepsilon - \mathbf{x}\|_1 \leq \varepsilon$, for accuracies $\varepsilon = \{10^{-1}, 10^{-2}, 10^{-3}, 10^{-4}\}$. To count the number of nonzeros in a vector $\mathbf{x}_\varepsilon$ for a particular accuracy ε, we first recall:

$$f_\alpha(\varepsilon) = \max_{\mathbf{P}} \max_s \min_{\mathbf{x}_\varepsilon} \text{nonzeros}(\mathbf{x}_\varepsilon) \text{ where } \|\mathbf{x}_\varepsilon - \mathbf{x}(\alpha, \mathbf{P}, s)\|_1 \leq \varepsilon.$$

Thus, we need to compute $\mathbf{x}_\varepsilon$ in a way that includes as many zeros as possible, subject to the constraint that the 1-norm of the difference between $\mathbf{x}_\varepsilon$ and $\mathbf{x}$ stays bounded by ε. The idea is to generate $\mathbf{x}_\varepsilon$ by computing $\mathbf{x}$ and deleting its smallest entries. The following steps illustrate our process to accomplish this:

- Compute the PageRank vector $\mathbf{x}$ with accuracy 10^{-12} via the power method.
- Sort $\mathbf{x}$ in ascending order.
- Determine the largest index j so that $(\sum_{k=1}^{j} \mathbf{x}_k) \leq \varepsilon$.
- Truncate these j entries to 0. Then $\mathbf{x}_\varepsilon$ contains $n - j$ nonzeros.

4.3   Testing the Theoretical Bound

To test the effectiveness of our theoretical bound in Theorem 1 we generate graphs with decay exponent $p = 0.95$, with different sizes $n = \{10^4, \ldots, 10^9\}$, and with maximum degree $d = n^{1/3}$ and minimum degree $\delta = 2$. Then we solve the seeded PageRank system, seeded on the node of maximum degree, with $\alpha = \{0.25, 0.5\}$.

For a more compact notation, let $\text{nonzeros}(\mathbf{x}_\varepsilon) = \text{nnz}(\mathbf{x}_\varepsilon)$ be the empirical number of nonzeros produced by our experiment. Figure 1 shows how $\text{nnz}(\mathbf{x}_\varepsilon)$ varies with n and α. Our theory gives a non-trivial bound once $n > 10^6$ for $\alpha = 0.25$ and $n > 10^8$ for $\alpha = 0.5$. These values of α (or near relatives) have been used in the literature [9,24]. That said, the theoretical bound stays far from the plot of the sparsity of the ε-approximate diffusion. Since the theoretical bound behaves poorly even on the extreme points of the parameter settings, we wished for a tighter, empirical bound.

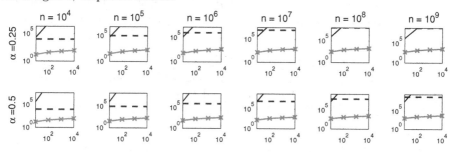

Fig. 1. A log-log plot of the quantity $\text{nnz}(\mathbf{x}_\varepsilon)$ versus $1/\varepsilon$ obtained for different experiments for $\alpha = \{0.25, 0.5\}$. We fix $p = 0.95$ for all plots, and run experiments on graphs of sizes $\{10^4, 10^5, 10^6, 10^7, 10^8, 10^9\}$. We choose $d = n^{1/3}$ and $\delta = 2$. The red dashed line represents a vector with all non-zeros present. The solid black line shows the bound predicted by Theorem 1. The blue curve shows the actual number of non-zeros found (Color figure online).

4.4   Empirical Non-zero Analysis

In this section, we develop a new bound that better predicts the scaling behavior of the number of nonzeros in $\mathbf{x}_\varepsilon$ as other parameters vary. We do this by studying the relationships among $\text{nnz}(\mathbf{x}_\varepsilon)$, ε, and p in a parametric study. Our goal is to find a function g where

$$\frac{\text{nnz}(\mathbf{x}_\varepsilon)}{d \log(d)} \text{ scales like } g(\alpha, \varepsilon, p). \qquad (8)$$

(The choice of $d \log(d)$ was inspired by our theoretical bound.) We first fix $n = 10^6$, $d = n^{1/2}$, and $p = 0.95$ and generate a graph as mentioned in Sect. 4.1. We then solve the PageRank problem and find the number of nonzeros for different ε values as mentioned in Sect. 4.2. We use $\alpha = \{0.25, 0.3, 0.5, 0.65, 0.85\}$ and count the number of nonzeros in the diffusion vector based on four accuracy settings,

$\varepsilon = \{10^{-1}, 10^{-2}, 10^{-3}, 10^{-4}\}$. We then generate a log-log plot of $\frac{\text{nnz}(\mathbf{x}_\varepsilon)}{d \log(d)}$ versus $1/\varepsilon$ for the different values of α. The outcome is illustrated in Fig. 2 (left).

From Fig. 2, we can see that as α increases, the values $\frac{\text{nnz}(\mathbf{x}_\varepsilon)}{d \log(d)}$ also increase, interestingly, nearly as a linear shift. Since we have seen that $\frac{\text{nnz}(\mathbf{x}_\varepsilon)}{d \log(d)}$ seems to vary inversely with $(1 - \alpha)$, we specialize the form of g as $g(\alpha, \varepsilon, p) = \frac{c_1}{(1-\alpha)} \cdot g_2(\varepsilon, p)$.

We similarly derive a relation between $\frac{\text{nnz}(\mathbf{x}_\varepsilon)}{d \log(d)}$ and p. Here, we fix $n = 10^6$ and $d = n^{1/2}$ then generate graphs with different decay exponents p, namely: $p = \{0.5, 0.75, 0.95\}$. We report the results in Fig. 2 (right). We can see that as the value of p increases, the curves $\frac{\text{nnz}(\mathbf{x}_\varepsilon)}{d \log(d)}$ appear to grow much more slowly. Furthermore, the difference between the curves becomes exponential as $1/\varepsilon$ increases. This leads us to think of the relation between p and $\frac{\text{nnz}(\mathbf{x}_\varepsilon)}{d \log(d)}$ as an exponential function in terms of $1/\varepsilon$. Also, since p and $\frac{\text{nnz}(\mathbf{x}_\varepsilon)}{d \log(d)}$ are inversely related, we consider $1/p$ rather than p. Therefore, we arrive at a relationship of the form:

$$g(\alpha, \varepsilon, p) = \frac{c_1}{(1-\alpha)} \left(\frac{1}{\varepsilon}\right)^{c_2/p^{c_3}}$$

for some constants c_1, c_2, c_3. After experimenting with the above bound, we found that the best results were achieved at $c_1 = 0.2, c_2 = 0.25, c_3 = 2$.

4.5   Results

The experimental scaling bound derived in Sect. 4.4 is now:

$$\text{nnz}(\mathbf{x}_\varepsilon) \leq d \log(d) \frac{0.2}{(1-\alpha)} \left(\frac{1}{\varepsilon}\right)^{1/(2p)^2}. \tag{9}$$

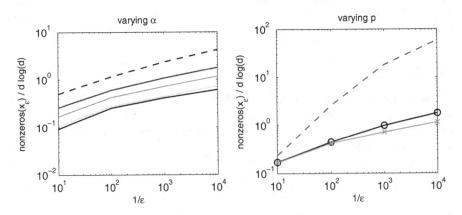

Fig. 2. Log-log plots of $\frac{\text{nnz}(\mathbf{x}_\varepsilon)}{d \log(d)}$ versus $1/\varepsilon$ obtained on graphs of size $n = 10^6$ with $d = n^{1/2}$ as α and p vary. At left, p is fixed to $p = 0.95$ and the black, green, blue, red, and dashed black curves represent $\frac{\text{nnz}(\mathbf{x}_\varepsilon)}{d \log(d)}$ for $\alpha = \{0.25, 0.3, 0.5, 0.65, 0.85\}$ respectively. At right, α is fixed to $\alpha = 0.5$ and the dashed blue, black, and green curves represent $\frac{\text{nnz}(\mathbf{x}_\varepsilon)}{d \log(d)}$ for $p = \{0.5, 0.75, 0.95\}$, respectively (Color figure online).

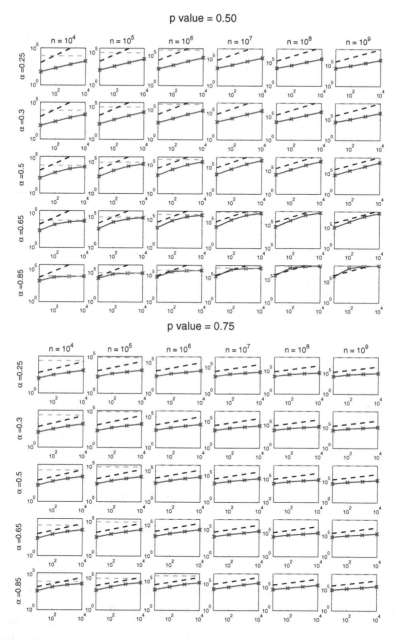

Fig. 3. Each sub-plot has x-axis representing $1/\varepsilon$ and y-axis representing the number of non-zeros present in a diffusion vector of 1-norm accuracy ε. The red dashed line represents a vector with all non-zeros present. The black dashed line shows our predicted bound (9). The blue curve shows the actual number of non-zeros found. As graphs get bigger (i.e. the fourth to sixth columns) the theoretical bound (black line) almost exactly predicts the locality of the ε-approximate diffusion (Color figure online).

p value = 0.95

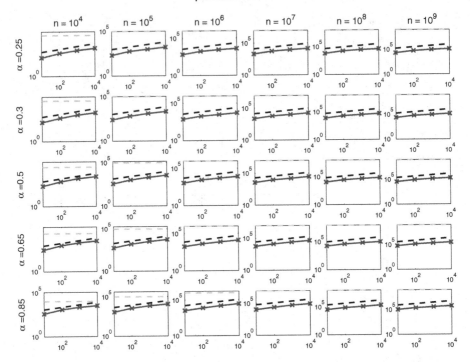

Fig. 4. Each sub-plot has x-axis representing $1/\varepsilon$ and y-axis representing the number of non-zeros present in a diffusion vector of 1-norm accuracy ε. The red dashed line represents a vector with all non-zeros present. The black dashed line shows our predicted bound (9). The blue curve shows the actual number of non-zeros found. As graphs get bigger (i.e. the fourth to sixth columns) the theoretical bound (black line) almost exactly predicts the locality of the ε-approximate diffusion (Color figure online).

In what follows, we demonstrate the effectiveness of this bound in describing the localization of seeded PageRank vectors computed with different values of α, on graphs with skewed degree sequences with varying decay exponents.

For each set of parameters (graph size n, $d = n^{1/2}$, decay exponent p, and PageRank constant α), the plots in Figs. 3 and 4 display the number of nonzeros needed to approximate a PageRank vector with 1-norm accuracy ε as a function of $1/\varepsilon$. The blue curve represents the actual number of nonzeros required in the ε-approximation. Each plot also has a black dashed line showing the prediction by our conjectured bound (9). We note that our conjectured bound appears to properly bound the empirical scaling in all of the plots well; although, it fails to provide a true bound for some.

5 Discussion

We have shown that seeded PageRank vectors, though not localized on all graphs, must behave locally on graphs with degree sequences that decay sufficiently rapidly. Our experiments show our theoretical bound to be terribly loose. In some sense this is to be expected as our algorithmic analysis is worst case. However, it isn't clear that any real-world graphs realize these worst-case scenarios. We thus plan to continue our study of simple graph models to identify empirical and theoretical localization bounds based on the parameters of the models. This will include a theoretical justification or revisitation of the empirically derived bound. It will also include new studies of Chung-Lu graphs as well as the Havel-Hakimi construction itself. Finally, we also plan to explore the impact of local clustering. Our conjecture is that this should exert a powerful localization effect beyond that due to the degree sequence.

One open question sparked by our work regards the relationship between localized solutions and constant or shrinking average distance in graphs. It is well known that social networks appear to have shrinking or constant effective diameters. Existing results in the theory of localization of functions of matrices imply that a precise *bound* on diameter would force delocalization as the graph grows. Although the localization theory says nothing about average distance or small effective diameters, it hints that the solutions would delocalize. However, solutions often localize nicely in real-world networks, and we wish to understand the origins of the empirical localization behavior more fully. Another open question regards whether localization is possible on graphs with a power-law degree distribution. Our current analysis is insufficient for this case.

References

1. Andersen, R., Chung, F., Lang, K.: Local graph partitioning using PageRank vectors. In: FOCS 2006 (2006)
2. Avrachenkov, K., Litvak, N., Sokol, M., Towsley, D.: Quick detection of nodes with large degrees. In: Bonato, A., Janssen, J. (eds.) WAW 2012. LNCS, vol. 7323, pp. 54–65. Springer, Heidelberg (2012)
3. Baeza-Yates, R., Boldi, P., Castillo, C.: Generalizing PageRank: damping functions for link-based ranking algorithms. In: SIGIR 2006, pp. 308–315 (2006)
4. Bayati, M., Kim, J., Saberi, A.: A sequential algorithm for generating random graphs. Algorithmica 58(4), 860–910 (2010)
5. Benzi, M., Razouk, N.: Decay bounds and O(n) algorithms for approximating functions of sparse matrices. ETNA 28, 16–39 (2007)
6. Benzi, M., Boito, P., Razouk, N.: Decay properties of spectral projectors with applications to electronic structure. SIAM Rev. 55(1), 3–64 (2013)
7. Berkhin, P.: Bookmark-coloring algorithm for personalized PageRank computing. Internet Math. 3(1), 41–62 (2007)
8. Bonchi, F., Esfandiar, P., Gleich, D.F., Greif, C., Lakshmanan, L.V.: Fast matrix computations for pairwise and columnwise commute times and Katz scores. Internet Math. 8(1–2), 73–112 (2012)
9. Chen, P., Xie, H., Maslov, S., Redner, S.: Finding scientific gems with Google pagerank algorithm. J. Informetrics 1(1), 8–15 (2007)
10. Chung, F.: The heat kernel as the PageRank of a graph. Proc. Natl. Acad. Sci. 104(50), 19735–19740 (2007)
11. Faloutsos, M., Faloutsos, P., Faloutsos, C.: On power-law relationships of the internet topology. In: ACM SIGCOMM Computer Communication Review (1999)

12. Freschi, V.: Protein function prediction from interaction networks using a random walk ranking algorithm. In: BIBE, pp. 42–48 (2007)
13. Ghosh, R., Teng, S.-H., Lerman, K., Yan, X.: The interplay between dynamics and networks: centrality, communities, and cheeger inequality, pp. 1406–1415 (2014)
14. Gleich, D.F., Kloster, K.: Sublinear column-wise actions of the matrix exponential on social networks. Internet Math. 11(4–5), 352–384 (2015)
15. Gori, M., Pucci, A.: ItemRank: a random-walk based scoring algorithm for recommender engines. In: IJCAI, pp. 2766–2771 (2007)
16. Huberman, B.A., Pirolli, P.L.T., Pitkow, J.E., Lukose, R.M.: Strong regularities in World Wide Web surfing. Science 280(5360), 95–97 (1998)
17. Jain, A., Pantel, P.: Factrank: random walks on a web of facts. In: COLING, pp. 501–509 (2010)
18. Jeh, G., Widom, J.: Scaling personalized web search. In: WWW, pp. 271–279 (2003)
19. Kloster, K., Gleich, D.F.: Heat kernel based community detection. In: KDD, pp. 1386–1395 (2014)
20. McSherry, F.: A uniform approach to accelerated PageRank computation. In: WWW, pp. 575–582 (2005)
21. Morrison, J.L., Breitling, R., Higham, D.J., Gilbert, D.R.: Generank: using search engine technology for the analysis of microarray experiments. BMC Bioinformatics 6(1), 233 (2005)
22. Nie, Z., Zhang, Y., Wen, J.R., Ma, W.Y.: Object-level ranking: bringing order to web objects. In: WWW, pp. 567–574 (2005)
23. Page, L., Brin, S., Motwani, R., Winograd, T.: The PageRank citation ranking: bringing order to the web. Technical Report 1999–66, Stanford University (1999)
24. Winter, C., Kristiansen, G., Kersting, S., Roy, J., Aust, D., Knsel, T., Rmmele, P., Jahnke, B., Hentrich, V., Rckert, F., Niedergethmann, M., Weichert, W., Bahra, M., Schlitt, H.J., Settmacher, U., Friess, H., Bchler, M., Saeger, H.D., Schroeder, M., Pilarsky, C., Grtzmann, R.: Google goes cancer: improving outcome prediction for cancer patients by network-based ranking of marker genes. PLoS Comput. Biol. 8(5), e1002511 (2012)

Author Index

Printed in the United States
By Bookmasters